THE GORODOMLYA ISLAND PROJECT

I dedicate this book to the many people like my father whose lives and personalities were pulled out of shape by the Cold War.

THE GORODOMLYA ISLAND PROJECT

THE INSIDE STORY OF HOW THE SOVIETS DEVELOPED ROCKET TECHNOLOGY

AMANDA VICKERS

Pen & Sword
MILITARY
AN IMPRINT OF PEN & SWORD BOOKS LTD.
YORKSHIRE – PHILADELPHIA

First published in Great Britain in 2025 by
Pen & Sword Military
an imprint of
Pen & Sword Books Ltd
Yorkshire – Philadelphia

Copyright © Amanda Vickers, 2025

ISBN 978 1 03610 8 823

The right of Amanda Vickers to be identified as the author of this work has been asserted by her in accordance with the Copyright, Designs and Patents Act 1988.

A CIP catalogue record for this book is available from the British Library.

All rights reserved. No part of this book may be reproduced, transmitted, downloaded, decompiled or reverse engineered in any form or by any means, electronic or mechanical including photocopying, recording or by any information storage and retrieval system, without permission from the Publisher in writing. No part of this book may be used or reproduced in any manner for the purpose of training artificial intelligence technologies or systems.

Typeset in INDIA by IMPEC eSolutions
Printed and bound in the UK by CPI Group (UK) Ltd, Croydon, CR0 4YY

The Publisher's authorised representative in the EU for product safety is Authorised Rep Compliance Ltd., Ground Floor, 71 Lower Baggot Street, Dublin D02 P593, Ireland.
www.arccompliance.com

For a complete list of Pen & Sword titles please contact:

PEN & SWORD BOOKS LIMITED
George House, Units 12 & 13, Beevor Street, Off Pontefract Road,
Barnsley, S71 1HN, UK
E-mail: enquiries@pen-and-sword.co.uk
Website: www.pen-and-sword.co.uk

or

PEN AND SWORD BOOKS
1950 Lawrence Rd, Havertown, PA 19083, USA
E-mail: Uspen-and-sword@casematepublishers.com
Website: www.penandswordbooks.com

Contents

List of Illustrations		vii
Foreword		ix
Author's Note		x
Prologue		xi
Chapter 1	Biographical Notes	1
	Percy and Edith	1
	Schooldays and Flitting	3
	Violence, Resentment and Anger	4
	A Good Brain	5
	Bristol Aeroplane Company	6
Chapter 2	Joining the RAF	7
	Basic Training	7
	Mission to the Fraunhofer Institute, 1946	7
	International Gathering of Scientists	25
	Jazz Party	41
	RAF Changi	53
Chapter 3	Air Scientific Research Unit, Obernkirchen, 1951	56
	Eric and Marianna	56
	Signals Intelligence	64
	Paddy's Antenna-Turning Mechanism	79
	Home Rescue Mission	81
	Abduction from Bückeburg Airfield	86
	The Brocken	89

Chapter 4	Gorodomlya Island	93
	Helmut and Irmgard	93
	N11-88, Branch 1, 1952	102
	Secret Radios	105
	Tasks and Inspections	109
	Transmitting to the British and Americans	110
	Skating, Tennis, Boat Trips and Concerts	113
	Finnish Christmas Concert	116
Chapter 5	Escape from Gorodomlya, Spring 1953	119
	Walter	119
	The Inspection	124
	The Big Freeze	125
	Escape Plans	132
	Escape to Finland, 1953	149
	Vaalimaa Border Control	154
	Helsinki Bus Station and the KGB	161
Chapter 6	Back to the UK, 1953	171
	Coronation Day	171
	RAF Gatow, Berlin	173
	Civilian Life, 1960	181
	Civilian Wife	198
	City 40, and GCHQ	209
Chapter 7	Escape to Canada	216
	Cold War Marriage	216

Epilogue	218
List of People in the Gorodomlya Island Project	219
List of Places in the Gorodomlya Island Project	224
Acknowledgements	227
Endnotes	229
Sources	238
Index	240

List of Illustrations

1. Ferry connecting Gorodomlya Island with Ostashkov. (*Anatoly Zak, 2002*)
2. My father and me in Hasfield, Gloucestershire, about 1970. (*Author's collection*)
3. Interior of the workshop on Gorodomlya Island, 1952. (*Werner Albring*)
4. Barbed wire surrounding Gorodomlya Island, strands 20cm apart. (*Georgy Karlov*)
5. Bruce Neville Cox (Cocky) with family members, 1958. (*B.N. Cox*)
6. Eric Ackermann as a young man at the Telecommunications Research Establishment, Swanage, UK. (*David Haysom*)
7. RAF Wilmslow, where B.N. Cox did his basic training. It was closed in 1962. (*Forces War Records (RAF)*)
8. Bedford truck, 1939–45. (*Imperial War Museums*)
9. The London Blitz: damage caused by V-2 rocket attacks in Britain, 1945. (*Wikimedia Commons*)
10. V-2 rocket launch from Peenemünde, Germany, 1942. (*Wikimedia Commons*)
11. In 1938, Korolev was sent to a gulag by Stalin in the 'Great Terror', for obstructing weapon development work. Paroled in 1944, this photograph shows a still malnourished Korolev in 1946. (*Boris Chertok*)
12. Sergei Pavlovich Korolev, mugshot taken by the NKVD (KGB), 1938. (*Natalya Koroleva*)
13. The barrack block, RAF Changi, 1948. (*B.N. Cox*)
14. RAF Obernkirchen in 1958, looking towards the entrance. B.N. Cox says in 1951, the facilities were 'primitive'. (*G.M. Stewart*)

15. The Brocken tower in 1951, before the iconic red-and-white transmitter was constructed. (*R. Demuth*)
16. Helmut Gröttrup in 1958. (*Ursula Gröttrup*)
17. Branch 1 workshop on Gorodomlya Island, 1952. (*TsIIMash*)
18. Rocket stand 2 on Gorodomlya Island, 1950. (*TsIIMash*)
19. Helmut and Irmgard Gröttrup, boat trip, 1950. (*Ursula Gröttrup*)
20. Inertial guidance system, produced 1959–62 for the US Thor IRBM (Intermediate Range Ballistic Missile), showing gyroscopes. (*US Air Force*)
21. Electronic inertial guidance system for project SPIRE. Developed by MIT from Hölzer's system in the early 1950s. (*Sanjay Acharya*)
22. Sketch of the G-4 (R-14) design concept by Konrad Toebe, who was a member of Helmut Gröttrup's team on Gorodomlya Island from 1946 to 1952. (*SchmiAlf, Wikimedia Commons*)
23. Zoom call in Covid lockdown, 2021. The author, brother David and father, B.N. Cox, aged 98. (*Author archive*)

Foreword
by Bruce Neville Cox

My daughter has taken some creative liberties with this account of a time in my life in the early years of the Cold War, when real war between Russia and the West seemed likely.

At that time, the scandal of the 'atomic spies' of the Manhattan Project, such as Klaus Fuchs, who illicitly passed information about nuclear weapons production to the Soviet Union, was still reverberating in the security services of the US and UK. Senator McCarthy's advisors Cohn and Schine visited the UK security services, looking for communists who might sympathise with the Soviet Union and therefore be motivated to pass on nuclear secrets.

Prior to Cohn and Schine's visit, security clearance in the UK was fairly casual, often amounting to a look to see if a person had a police record, or membership of the Communist Party. I had to be rescreened for work with Eric Ackermann, although I had previously been screened for the Air Ministry, and this took months.

Gorodomlya Island was just one tiny piece of a massive jigsaw of laboratories, factories, mines and design offices, all established for the purpose of the USSR to compete with America to produce a rocket-carried nuclear warhead capable of inflicting unsurvivable damage on other countries. That was the 'project'; what happened on the island was a small cog in it.

Gorodomlya scientists with sensitive information were held back (from release) for over a year before being debriefed by GCHQ and the CIA.

Scientists everywhere were aware of the enormity of responsibility they had in their hands. Slowing and sabotaging developments was commonplace. That was my strategy and I believe it was Korolev's. No sane person wants these extreme weapons to be deployed against humans.

It all comes down, in the end, to sanity.

Author's Note

This is a non-fiction book, based on one person's memories of the events of eighty years ago, as he neared his hundredth year. It is based, too, on the sources I have used to corroborate and expand my understanding.

My aim throughout has been objectivity, but in the process I have discovered there is no such thing as objective truth when it comes to politics. The Cold War, and the technological developments that sprang from it, were, and still are, highly polarised, politicised and controversial. Every source contains bias, either towards the Soviet worldview, where every discovery is ascribed to Russian science, or towards Western Europe and the USA.

Dates, places, names, motivations and achievements all enter a quicksand of subjectivity. There will be readers keen to point out inconsistencies shown by their own research, and although I am interested to hear from them, the next reader will claim a different version.

As I narrate the story of a part of my father's life, I use 'injections of creativity' in the form of dialogue to bring personalities to life. These serve the narrative by staying close to known realities, and are based on deduction and clues. Through a highly contested history on the famously impenetrable subject of rocket science, I have sought objectivity, but I cannot claim truth.

Important truths, such as the futility of rule by fear, the need for mutual trust, and the strenuous and risky nature of peace seeking, emerge from the narrative.

On the day my father died in Canada, his granddaughter, a nuclear physicist, ran the 2024 London Marathon. I could hardly hear the phone call with the news because of the crowds lining the streets. Life, genes, and scientific endeavour, continuing.

Amanda Vickers

Prologue

It's 1972. Two policemen are at the kitchen table with my mother when I come home from school. I push my key into the lock as usual, but the door is already open, my brothers shooed upstairs. I stand in the hallway, hidden by hanging coats and schoolbags, listening.

My father had not come home from work two days before. He had set off, grim and silent, no goodbyes ever expected. When my mother, who had his evening meal ready at six, phoned his company to ask where he was, they had been evasive.

Today, apparently, the personnel manager had rung my mother. They had found his company car, at a long stay car park in an airport, with a ticket for the maximum possible stay. There was more. My father's secretary had disappeared from work as well. Her husband had reported her missing. The police were trying to trace her, as a missing person, but my father's company were more concerned about getting the company car returned.

We finally ate the meal in tense silence. I don't know why I still expected his car to pull up in the drive, and was upset when it didn't, but I was only 9. His footsteps did not come up the stairs, even though I was listening out for them.

The airport eventually provided the passenger lists of every flight that had departed in the last couple of days. Neither his name nor his secretary's were on the lists.

One policeman in particular became a fixture in our kitchen in the months that followed. He had interviewed the work colleagues of the two disappeared people and discovered their affair was common knowledge. Her husband had a history of domestic violence against her, which was known to the police. He had been so quick to report her missing because he suspected she was planning something. He said he didn't want to be left alone with their three children.

'I'm also left with three children,' said my mother.

'We're doing all we can to find them,' said the policeman, 'but the trail has gone cold.'

There was a breakthrough. My mother found a London A–Z street map with red crosses marked in pencil on a series of guesthouses in the area near the airport. The policeman contacted each guesthouse with descriptions of the couple. A false name, Mr and Mrs Cook, had appeared on several registers over the last couple of months. The airport had stopped checking passenger identities. When the policeman insisted on one final passenger list check, their real names came up, sitting together, on a one-way flight to Canada. But the list was dated two weeks before. They had escaped.

'Blimey, he's slipped through our fingers like a spy,' said the policeman, banging his forehead with the palm of his hand.

'I suppose I somehow have to get a divorce,' said my mother.

∗ ∗ ∗

Fifty years later, in 2021, the country is locked down because of a pandemic.

My younger brother, now in his fifties, tells me that he has been talking over Zoom with our 98-year-old father, and that he is living in a care home in Canada. He still lives with his wife, the secretary he disappeared with when we were children. She is blind now. They live in the care home for her sake as my father is as fit as a flea. I am ambivalent about this news. He had left my mother and us, his three children, in near poverty, without a backward glance. His own mother, our Nanny Edie, had died years ago. When his sister tracked him down to give him the news of her death, he didn't return from Canada for her funeral despite apparently caring more for her than for any other human being.

I'm not sure if I want to talk with him, and I wrestle with indecision for months. How will I be with him? He chose to miss my entire life. How could I pick up the threads of being his daughter when there wasn't a warm remembered past?

∗ ∗ ∗

I remember my mother putting me, aged 3, on his knee. The ridges of corduroy unfamiliar. No arms came around me and I jumped down, confused. I recall a photograph on top of a gate, when we lived briefly in the Gloucestershire countryside. Me, aged 6, opposite my father. My mother telling me to look at him and smile. I smiled as directed. He managed it too. A dishonest picture, even then. Just before he disappeared, when I was 9, I asked him for a try with his headphones, always on his ears and another form of separation. He placed them on my head without turning the volume down. A blast of classical music so loud it hurt. I pulled them off.

These memories do not warrant an adult relationship. I am angry at the depth of his neglect. All my achievements, and my happy family life, are my own. I don't want to hear 'You got that from me'.

* * *

My brother is still talking. 'No, there's a whole story, it's not very coherent. He tried to explain how he got pulled into spying, when he certainly didn't want to take risks.'

'No one becomes a spy by accident,' I say, though I know nothing whatsoever about it.

'He did. He was targeted by the Russians ... there was an island in Russia where ... people were abducted ...'

'Sounds mad,' I say, hardly interested.

Dad had been encouraged to write his life story by care home staff, and he was doing it, right back to his childhood. He managed to qualify as an engineer, even though his father tried to stop him getting an education. He says he passed on 'scientific secrets' in the Cold War.

My brother was most upset by how violent his father, our grandfather, had been. A seaman – who hit his wife. 'Hit his wife? Our Nanny Edie?' I am incredulous. Nannie Edie was the sweetest person.

'Yes. When Dad was old enough, earning enough, he moved our Nannie Edie to a flat of her own, safe from her abusive husband.'

I concede that I'll read his biography, written only for family members.

'Look, think about it, Dad is 98 and may not always be compos mentis. Why don't you join our next Zoom call? You don't have to speak. I'll be there to do the talking.'

* * *

With my brother leading the call, I take time to observe this elderly man. Obsession with the weather, a characteristic lack of warmth, no apparent regret about missing our lives. The first attempts were difficult. Only a lack of distractions in lockdown kept them limping on.

The calls became peppered with odd rememberings. A man, maybe a friend, called Eric Ackermann, who wasn't German, some sort of boss in the RAF. A long journey to Austria after the war, to remove the guts from a top-secret laboratory. Meeting someone there, a Russian, who turned out to be famous for space exploration, name forgotten. The Russian's sidekick's name, 'Mitya', still fresh.

'What about the island, Dad?' I asked. 'What was it called?'

'You're not supposed to know anything about that,' he countered.

'It doesn't matter now; you're not bound by the Official Secrets Act anymore. I'm just interested. You said you'd been taken there against your will. What happened there, Dad?'

'I can't. I've never told anybody.'

'Was it a prison camp? Were you a prisoner on this island?'

'It was sort of a prison camp. A prison camp for rocket scientists.'

'Were you a rocket scientist, Dad?'

'No. They thought I was, but I was an electronics engineer. I was the only Brit on the island. Everyone else was German or Russian. The Germans thought I was Russian, the Russians thought I was German! I just kept quiet, mostly.'

'Which Germans?'

'Helmut Gröttrup was the boss. I stayed with him, his wife Irmgard, and their kids. They were always at each other's throats. I liked her, very much. She made some headphones for my radio, using her furry earmuffs! I really shouldn't have made that radio. I could've got killed for having it. One of the

Germans, Walter, had a homemade radio that he used in the same way I did, sending information back to the Brits and the Americans. He got sent to a plutonium plant. I feel sad about that still.'

'Dad, where was this island? Can you remember the name?'

'Oh yes, I know the name.'

'If you tell me it, I'll look it up on the internet, and together we can find out more.'

'Alright. Don't tell anyone else, though. It's Gorodomlya Island.'

***

That was how my lockdown project started, under the government edict 'Stay at home'.

There is a fair bit of information about Gorodomlya Island on the internet. Information about Eric Ackermann and his wartime and post-war work, and the fact he reported to R.V. Jones, a government physicist who had Churchill's ear. About the German development of V-2 rockets at Peenemünde[1] during the Second World War; about Operation Paperclip when those scientists were offered rocket-building jobs in America; and the Soviet response to that, by pulling off a mass abduction of rocket and nuclear scientists to locations around Moscow, including Gorodomlya Island.

My father had a narrow perspective on his own experience and contribution. He was trying to overcome a lifetime's silence. When I asked him why he couldn't talk more freely, he told me that he used to be so talkative, perceptive and annoying that his RAF friends nicknamed him 'Cocky'.

His name is Cox, the same as my pre-marriage name. His experiences in the Cold War and the need for secrecy for more than a decade of his life, he says, rendered him 'a bit boring'.

'I'm writing all this down on index cards and trying to file it in chronological order,' I tell him.

'Good luck with that,' he chuckles.

***

It was my own scepticism in relation to the odd details wrung from him so painstakingly that got me checking and researching the events that my father wrote and told me about. How he became involved with the rocket scientists on Gorodomlya Island. How he might have made an insignificant contribution to how the rocket was guided, the one that eventually launched the first satellite into space, in 1957. The one intended by Soviet Russia to carry a nuclear warhead as far as London or New York.

I found external accounts to support some of his memories. I cross-checked the rocket science with Leeds University Rocketry Association. I joined the British Library to access books and records; I looked up my father's military history. I was trying to disprove, trying to catch him out, trying not to care, trying not to do him the honour of understanding more about his life.

I ended up talking with my father regularly on Zoom. I ended up writing this book.

Chapter 1

Biographical Notes

Percy and Edith

Our family tree begins with a John Cock, yeoman farmer of North Wootton, Somerset, born in 1718. A child of his, Phillip Cock, married Amelia. They had ten children, including Phillip Cock, born 1774, who married Ann Norman. Phillip and Ann had eight children, including John, born 1812, who married Charlotte Tucker, born 1815. About this time, their surname starts to be spelled 'Cox'.

John and Charlotte had nine children, including Henry Cox, born 1853, who became the village tailor in Painswick, Gloucestershire, an historic wool and weavers' town of around 5,000 population. Now it is known as 'Queen of the Cotswolds', but at that time, the majority of people living in Painswick were poor. Henry married Elizabeth Hayman, and their seven surviving children included five boys, one of whom was Percival (Percy Cox). According to family stories, these five boys terrorised Painswick with their wildness, but the two girls, Kate and Charlotte, were well liked.

Percy Frederick Edgar Norman Cox, born 1891, is my father's father, called Iggy in life and in the book, a nickname given by my Great-grandma Becky, Edith's mother, that is some reference to 'Father Ignatius, The Great I Am'. Percy married Edith Maude Wilkins, my 'Nanny Edie', on 7 March 1915. Edith's parents were William and Rebecca Wilkins of Wales. Percy and Edith had nine children, born 1916 to 1938; my father, Bruce Neville Cox, was their fourth child and third son, born 19 January 1925. Only six of their children survived infancy.

Percy was an able seaman, mainly based in Bristol ports, though he implied, through projecting a well-heeled and well-dressed charming persona with pink gin in hand, that he'd been a naval officer in the First World War. He did serve on a destroyer in the Battle of Dogger Bank in January 1915. Percy (Iggy) always had a car, despite his family living in abject poverty and having to flit in the night at least eleven times because of unpaid rent. He was a serial philanderer and he couldn't keep a job because he would succumb to fiddling and fraud, and openly drink the proceeds in the nearest pub. He had a fear of any of his children succeeding in life, doing all he could to spoil their every opportunity. He was a violent, angry man at home, though a charmer outside of it. He may have killed my father's brother Bernard.

I remember a visit as a child to a dark flat in Bristol. Nanny Edie opened the door and begged me and my brothers to keep quiet because my grandfather, Iggy, was in the back room. I accidentally went into the room where Iggy sat in the dark in his armchair. He roared with anger at the sight of me, turning me into a gibbering wreck for the rest of the visit.

My father won a scholarship to study at a prestigious school, but:

Father [Iggy] true to form, pulled me out of school at Easter '39, just after I was 14. I was taken on at Filton Works [Bristol Aeroplane Company] for twelve shillings and sixpence per week. The odd two shillings and tuppence was mine. The rest went to mother. As she still had difficulty making ends meet, it was clear that Iggy docked my wage from her housekeeping money. Even so, I managed to buy a second-hand bicycle, and joined the ARP as a bicycle messenger.

Just one of several similar accounts in my father's life story goes like this:

Father got fired from Stone & Jenkins for fraud. He would take an order for parts from a garage owner, offer a lower price if the order was paid for in cash, not supply the parts and drink the money in the nearest pub. Stone & Jenkins was inundated with angry demands from several garages for the spares they'd paid for. The firm nearly went under but managed to weather the debacle. If they'd chosen to prosecute, he'd have gone to prison and later on, when John and I were mulling over 'the old days', we decided that we'd have been better off if he had!

All the siblings encountered people who saw their potential and were willing to help them, perhaps understanding the burden of being a child of a known petty crook. This next account shows how the same company that nearly went under still believed in my father's brother Basil, known as John:

> Mr Jenkins, the other partner at Stone's, took a shine to John and used to augment his wages from his own pocket. He paid for John to learn to drive and also his fees for night classes to study engineering. John was able to buy a snazzy 'racing bike' – a Sun Wasp – and joined the Bristol South Cycling Club. From then on, he was always away most of the weekend, keeping out of father's way. We all avoided Iggy as much as possible.

Schooldays and Flitting

My father's mother, Edith, was the parent who believed in getting the best education possible for her children, but she was up against continual uprooting and moving to all sorts of barely planned places because of 'the need to stay one step ahead of the bailiffs'. From the age of 4 to 10, my father moved with his family at least eleven times, mostly within Bristol but also to the Isle of Man and Liverpool. Edith taught him to read and tie his shoelaces as a prerequisite for starting school, but much of his education was picked up through childhood jobs. When my father, known as Neville, was 8 years old, Iggy had a battery charging business, where customers would bring in their accumulators for charging. Young Neville would carry two charged sulphuric acid batteries through the streets to the owners' houses, not daring to drop the heavy, dangerous things. He got to know a group of RAF children, boys who formed a cricket team, when he lived at Gilda Crescent, near Whitchurch Airfield. The RAF boys attended Filton School, and through them Neville was able to play on a proper cricket pitch and allowed to use good sports equipment.

At age 10, Neville sat the scholarship exam and was offered a place at Queen Elizabeth's Hospital School, a boarding school that would have saved him from his home life. He says, 'I was agog to go but father wouldn't sign the papers as it meant I'd have to stay until I was 16.' Instead, he went to

South Bristol Central School. It had an engineering and science bias, with workshops for wood and metalworking. At age 14, his father pulled him out of this school to take up a job locally at the Filton Works of Bristol Aeroplane Company.

A formative experience for Neville was at age 13, when Whitchurch Airfield hosted a flypast of the military air power possessed by the UK at the time. Many bombers were biplanes, but the Handley Page Heyford was supposed to be the latest thing in operational bombers. According to Neville, on the day of the flypast the plane 'trailed the pack', taking the whole length of the airfield to take off, and over an hour to climb to 5,000 feet. The crew were seen to parachute out before the aircraft plunged into the Bristol Channel. He writes, 'We didn't know what was going on in the background that eventually resulted in some decent aircraft,' but when he started work at Bristol Aeroplane Company, he was very interested in building better aeroplanes.

Violence, Resentment and Anger

The trauma of witnessing his father's mistreatment of his mother – the financial abuse, physical abuse, and the way he scuppered everything she wanted for her children whilst insisting on being looked after as if it was her duty to serve him – never left Neville.

'I won't forgive him the waste he made of his own talents and of the lost opportunities of his children,' he says, pointing out that Iggy was 'nasty to the end: forcing Mother out to shop in her eighty-sixth year. She collapsed in the street from a stroke and died two years later, never regaining the ability to speak or walk.'

Neville deplored violence against women and children, and hated aggression in any form. The tyrannical nature of the Nazi regime and Stalin's regime have obvious parallels. To my mind, Iggy was made of the same stuff as Stalin, except that Stalin had a whole nation to terrorise, not just a family. Anger at brutish, bullying behaviour often turns inward in children, who either display their emotions and are considered naughty, or internalise overwhelming feelings of being victimised and become watchful, shadowy … waiting for their moment to take revenge.

My father wasn't guilty of physical abuse when he was married to my mother for ten years. Instead, he was neglectful of her, and of us, his children. He had no consideration for the family's need for stability or for my mother's work as a teacher. We moved house often. He was silent, obsessed with mechanical and electrical things such as his cars. His love for classical music remained uncommunicated – his headphones saw to that. He had no expectation that enjoyment could come from knowing his children. It is nothing short of a miracle that two out of three of his children know him now.

A Good Brain

A recurring theme in my father's memoir is how outsiders to the family recognised potential in him. From teachers at school to employers, people refrained from writing off the entire family. Sometimes their practical help – a raincoat for a child with a long walk to school, a pencil case, or money for a pair of shoes – made a difference. Despite the dysfunction and upheaval of his home life, Neville got the idea that he had a good brain, and that he needed to nurture it.

Neville was born amid the economic depression of the 1920s and 30s, when, due to the First World War, there were massive shortages in qualified craftspeople, scientists, engineers and technicians. Like many others, he made use of night schools for adult education in order to get qualifications he could not gain in any other way. Attending in the evenings after work, he again found that he was well supported. The qualifications were rigorous but lacked official equality with university degrees. Education for the working classes was becoming a major socialist ideal. At colleges like these, many working-class men discovered they had a good brain. It must have taken a certain amount of confidence and conviction for them to believe they had the ability to take and pass these courses. Self-belief is a fragile thing, but the encouragement of a few enlightened people in Neville's young life sparked him into having faith in his abilities.

When Neville joined the RAF in 1946, the service was casting off its massively damaging assumption that only a university degree indicated intelligence. It was starting to value not just the seemingly undiscovered quality of women's brains but also the remarkable ability of non-university people to gain and apply knowledge, solve problems, and accomplish work equal to or exceeding that of people with privileged educational backgrounds.

Bristol Aeroplane Company

Founded in 1910, the Bristol Aeroplane Company's first premises were two former tram sheds at Filton – once a village but at this time becoming an urban overspill of Bristol. From 1929, No. 50 (City of Bristol) Squadron was based at RAF Filton. The UK government used the Filton Works extensively to design and produce the warplanes needed for the war effort. My father worked in the technical drawing office for about five years, from age 14 to 19, while also attending night school almost every evening. Here he was increasingly involved in designing interiors, engines and instrument panels for warplanes to be used in the Second World War.

On 25 September 1940, when Neville was working there aged 15, the Luftwaffe[1] raided Filton, aiming for the aircraft works, and killed at least 143 employees. It was just before midday, and the most casualties occurred when air-raid shelters full of people were hit. The churchyard of St Peters at Filton has a memorial to those who died in the raid.

In 1959, the company merged with several major British aircraft companies to form British Aircraft Corporation, now BAE Systems, which still has a presence at Filton.

Chapter 2

Joining the RAF

Basic Training

From age 16, Neville kept trying to join the RAF, but his role in the design of warplanes at Filton was needed and his applications were blocked by his employer. When his home life became unbearable, he left Bristol to live with his brother John[1] in London.

When he was finally accepted in January 1946, there were only two vacancies in the whole of the RAF. Because the war had ended, he applied for a civilian role. He does have a military number, and he did his basic training at RAF Wilmslow in Cheshire, but here he was allowed to 'lounge about in the offices' while everyone else was square-bashing. He says: 'I never did a parade, guard duty or anything involving marching' because it was considered that his trade didn't require it. Instead, he 'produced posters for sergeants' mess dances and the occasional statistical chart'. He enjoyed the closeness of Wilmslow to Manchester, where he used to go to see the Hallé Orchestra playing on Sunday afternoons.

Being appointed as a draughtsman for the Fraunhofer expedition was to him like applying for any civilian job. The RAF trappings of uniform, jargon and hierarchy never became ingrained. He quickly gained a huge respect for Eric Ackermann and the engineers he worked with in the RAF, but he was never 'one of the lads'.

Mission to the Fraunhofer Institute,[2] 1946

The Fraunhofer expedition, or mission, was undertaken so that the British government could get access to parts, documents and instruments associated with the development of Germany's wartime 'wonder weapon', the V-2 rocket. The idea was to relocate the entire laboratory and its remaining staff from Austria into the British occupation zone in Germany. The mission was arguably Eric's initiative, with support from R.V. Jones[3] and Winston Churchill, with Roy Piggott leading as an established government scientist.

* * *

For Neville, draughtsman for the mission, the main value was in meeting Eric Ackermann. Eric was setting up his Air Scientific Research Unit (ASRU) in Obernkirchen, supported by R.V. Jones.

It was difficult to see why this mission mattered to this narrative at first, largely because my father had totally forgotten the name of one of the Russian visitors, Sergei Korolev,[4] who went on to lead the Soviet space programme. Korolev at this time wasn't the impressive leader he was to become, and his name remained unknown outside of the USSR[5] until after his death.

On this mission, Neville met with Russians from Bleicherode,[6] the location of the RABE rocket-production works, in central Germany.[7] These Russians were combing out research bases[8] previously staffed by scientists until recently employed by the Nazi regime. The group of Russians were under pressure to identify 'the specialists' behind the V-2 rocket. Boris Chertok, a Soviet control systems engineer who worked under Korolev, states in his book *Rockets and People*, Volume 1: 'We agreed that we would set up our own independent intelligence service. The primary mission of this group will be to search for authentic missile specialists and entice them, or even abduct them, from the American zone.'

* * *

My father insists he was small fry when compared to other scientists. He was only 20. He wasn't a rocket scientist.[9] He had just a layman's knowledge of the V-2, and had only just joined the RAF. Because of his insistence on not knowing anywhere near enough to be of value, the question of why, five to six years later, he was targeted by the Russians is something not immediately understandable. My hypothesis is that his own cockiness on this mission, combined with being in the wrong place at the wrong time, brought him to their attention. He was barely aware of the underlying politics, instead being overly concerned with projecting his knowledge and making accurate technical drawings.

* * *

The following narrative creatively reconstructs the expedition as remembered by my father and corroborated by Nick Ackermann, second son of Eric, who claims he can 'hear his father's authentic voice'.

A long line of seventy battered British Army trucks lurched and rumbled through the night, sinuously connected like the segments of a caterpillar. Moving slowly because of the icy, bomb-damaged roads, each kept to a distance of 3 feet between itself and the one in front. Five motorcycle outriders flanked the convoy, three in front and two at the back. Drivers fought to stay alert, against the blackness of the night and the soporific swaying of overheated cabs. The convoy cut a swathe through hundreds of miles of snowbound German-Austrian countryside.

It was February 1946, and the Second World War in Europe had ended in May the previous year. Now it was deepest winter, and the planning for this trip had been done quickly, wearily, by people at the Air Ministry who were looking forward to getting back to their pre-war careers. Most drivers were young, recent recruits who hadn't seen any fighting but had done their basic training. Others were veteran pilots injured in battle. Some rookie drivers had just passed their military driving test the day before, hurriedly pushed through the basics. A few had simply been asked if they could drive, and were issued on the spot with their military licence.

Permission for the mission had been cleared at the highest government levels, and Air Ministry intelligence expert Dr R.V. Jones was taking a personal interest. Jones was reporting on progress to Winston Churchill, currently active in the Air Ministry as Leader of the Opposition, having lost the July 1945 election. A few months before, R.V. Jones had met with one of his pilots and fellow scientists, Eric Ackermann, and was supporting Eric's plan to undertake this mission, against a tide of war-weariness and into a fragile post-war political situation. R.V. Jones had authorised an experienced military scientist, Roy Piggott,[10] to lead the mission, with Eric, to his great disappointment, only second-in-command.

Eric was 26, and was well known for his headstrong personality. He had a reputation for being a daring pilot as well as a scientist with first-hand experience with intercepting the 'beams' used by the Luftwaffe to guide their bombers. He was known as one of the least manageable officers in the Royal Air Force.

The seventy army-issue Bedford trucks, supplies and personnel, complete with mobile kitchen and motorcycle outriders, had materialised from the surfeit of redundant vehicles after the war. The convoy was one of the final things that many in the wartime cabinet had to think about. Everything was winding down, but, according to Eric, 'Some people in the Air Ministry still wanted their telephone to ring occasionally.'

In truck 38, somewhere in the middle of the convoy, Neville Cox, a 20-year-old Royal Air Force draftsman, with his new LAC (leading aircraftman) rank, was pulling worn-out blankets from behind the passenger seat he reluctantly occupied. His aim was to form a pad that would protect the backs of his legs from the broken, gutless seat. The young draftsman could drive, probably better than many of the hastily recruited drivers, but designated driver Eric Ackermann had insisted on the new recruit being his navigator. The younger man would only take over as driver if Eric needed to rest. Eric was too stubborn to give up his status as driver. He was hell-bent on keeping his eyes open for the duration of the journey.

There was absolutely no need for navigation for any but the first truck in the convoy, which was being driven by physicist Roy Piggott. Roy had an enormous dog-eared map spread out on the dashboard of his truck, which he referred to occasionally, never slowing down long enough to study it.

Eric didn't need a navigator, but he wanted to find out more about the young draftsman. He had asked for someone with technical drawing skills to be taken on for the mission. All he knew about the new recruit was that he came highly recommended by his previous employer, Bristol Aeroplane Company. Apparently, he had brought to his Air Ministry interview a portfolio of accurate diagrams of electronic aeronautical equipment, showing how it all fitted together.

Neville had applied for the RAF twice since turning 16, before being accepted, but his long-held dream of flying a warplane was stymied by the end of the war. He was unsure now what a career in the RAF would constitute. He had been looking forward to training as an RAF pilot and was dismayed to be tasked with yet more technical drawing. He had now to endure a journey as a passenger in a particularly old and decrepit truck. He had already asked several people what the purpose of the journey was, but questions were clearly unwelcome. People just got on with the task in hand. He knew a little about the mission from his interview, but his curiosity was unsatisfied.

Both men were wearing blue RAF uniforms. Eric's was worn and old, the trousers almost threadbare at the knees; but the younger man's was stiff and new, itchy on his neck, with a shiny leather belt that went round his waist almost twice, nipping in the trousers around his skinny form.

'That seat's shot,' said Eric. 'These trucks are all near the end of their useful lives. I'll be bloody jiggered if we all get there in these ancient relics without at least some of them breaking down.'

The younger man was still struggling with the seat and was looking in vain for any control mechanisms for it. The seat sloped towards the driver. He couldn't sit straight without being thrown towards Eric when the truck lurched. In the short time Neville had been around the older officers, he had already irked them with his bright-eyed curiosity, and he had picked up the nickname 'Cocky'.

'Some of the newer Bedfords have seat harnesses,' said Eric. 'Certainly, the new signals vans do. But not this one.'

Cocky piled the blankets ostentatiously high where the seat dipped, and jammed his leg on top of it all so that he was, at least in theory, more balanced.

'Hard luck, Cocky, you'll just have to harden up. You're in the RAF now, and people learn not to complain – if you can't put it right for yourself, no bugger else will do it for you.'

'Just let me know when you want to swap,' replied Cocky brightly. He would have been much happier in the driver's seat, even though it was far removed from the cockpit of his dreams. He was also keen to be more comfortable, and the driver's seat had been repaired with a bright yellow wad of new foam padding.

'You might do well to remember that I'm a flying officer,' said Eric. 'I'm your superior. The quicker you understand where you rank – as in, as leading aircraftman, LAC, you are at the very bottom – the better. What is it with you? You're not exactly deferential, are you?'

'Sorry, Sir.' Cocky was unused to RAF protocol, and was still working to understand the pecking order.

Eric was worried that he might have overstated his adherence to questions of rank. He wasn't particularly interested in rank, other than the fact that promotions brought higher pay. He certainly didn't want to spend this entire journey with someone he had frightened into silence. He tried again.

'I'm probably the most laid-back flying officer you will ever come across,' he said in a more friendly tone. 'And you having a lip on you doesn't matter here between the two of us in this cab. But don't make the mistake of opening your mouth in the presence of senior officers. They don't want your opinion and you could end up with something worse than a nickname.'

'I like my nickname,' said Cocky, cockily.

Eric took a moment to decide how he was going to respond. If he was too sharp with the prickly new recruit, there would be hundreds of miles of strained, dull quiet in front of them. 'Mine's "Acky",' he said with forced cheerfulness. 'My real name, "Ackermann", sounds German, because it is. I have some German ancestors.' He looked over at Cocky but there was no reaction to this revelation. Cocky's face was a picture of studied non-response.

Eric ploughed on. He was born on the Isle of Wight, brought up in London, and was most definitely British. Now the war was over, he was reclaiming 'Ackermann' as the name he was born with. He was considering dropping one 'n' to make it more English. His hope was that being of German descent wouldn't go against him in a country where, for most people, 'German' still equalled 'Nazi'.

The truck rumbled on for a good while as Cocky decided whether or not to risk speaking. Eventually, his curiosity won and he piped up with a direct question about the laboratory they were heading for, asking whether it used to be 'an actual Nazi one'.

At this point, Eric went silent. He hadn't been the one to tell Cocky that the purpose of the trip was to strip the contents of a key wartime laboratory.

'Who told you that?' he exploded. 'It's top-secret classified information! No one should be telling you things like that. Not even R.V. Jones would have told you, even if you have signed the Official Secrets Act. I'd like to know, who the hell told you?'

'Churchill. By accident. I accidentally asked Churchill.'

'What?' shouted Eric.

'Winston Churchill,' confirmed Cocky clearly, his eyes straight ahead.

'You're going to have to explain yourself,' barked Eric.

Cocky cleared his throat and spoke evenly and clearly. He wished that Eric would stop staring into his face. He was uncomfortably aware that Eric was looking for evidence of lies. Cocky explained how he had been waiting outside the interview room at the Air Ministry for over an hour without anything to read, not even a window to look out over London. He had walked a little way down the corridor, to the top of the stairs, just as Churchill[11] was coming up them with a few other men. He didn't want to run away; it would have looked suspicious. One of the men asked him what he was doing there. He had his security pass with him and was just getting it out of his pocket, when Churchill told this someone else not to worry because he was being interviewed for the 'Fraunhofer expedition'. He didn't know what that was, so he asked about it, and Churchill told him that it was about stripping an important laboratory in Austria of its experimental direction-finding apparatus. Then the group wished him luck and walked on.

'Bloody hell,' said Eric, whistling softly to himself. 'Perhaps it pays to be cocky sometimes.'

Cocky relaxed a little. He hadn't mentioned meeting Churchill in this way to anyone other than his brother John, with whom he was lodging in London, because it didn't seem significant.

'One of the men who interviewed me was Edward Appleton. So, I at least know who one government scientist is. Though I think I upset him.'

'You will have done,' said Eric. 'You will most certainly have upset Sir Edward Appleton.'

Cocky wondered wretchedly if Eric was as annoyed as he sounded. Eric was quiet for a long time and when he spoke again it was to underline that the whole mission, to move the entire Fraunhofer laboratory, its equipment, instruments, documents and remaining staff from the American zone and into the British zone, had been his idea. He explained how he wanted to gain as much new technical information about German air scientific developments and inventions as possible, so that R.V. Jones could distribute them to British universities and scientific institutes.

'It was your idea?'

'Yes, it was my bloody idea,' exploded Eric. 'My idea, my mission.'

'Don't the Americans want the laboratory? It's in their zone.'[12]

'They've already had first pick of the scientists involved,' said Eric. 'Haven't you heard of Operation Paperclip?'

'No,' said Cocky.

'Alright. Operation Paperclip started last year, at the end of 1945, when Truman, the American president, offered homes, well-paid scientific jobs, and a high standard of living to a whole group of the scientists from Peenemünde. Most of them couldn't believe their luck and they went to the United States with their families, pretty damn quick.'

Cocky hadn't seen this in the newspapers.

'Only the director, he's Doktor Dieminger,[13] – well, Appleton and Piggott call him "The Doc" – is still at the Fraunhofer, with about twelve of his closest colleagues. His life's work is there, and he wasn't interested in Truman's offer.'

Cocky asked about Dr Dieminger's 'life's work' and was plunged into a morass of scientific detail.[14] Eric explained that pre-war, the work undertaken at this particular laboratory was about researching the ionosphere. Cocky had never heard the word 'ionosphere'. It sounded like something to do with the Earth's atmosphere. He had a sketchy understanding that it had two layers[15] – one that started at sea level, going up about 60 miles, and another that went up much further, to the edge of space. He was dreading the moment when Eric would ask him what he thought the ionosphere was. He hated admitting gaps in his knowledge.

Luckily, Eric was in full flow. He explained that before the war, the laboratory was gaining an international reputation because there were some brilliant scientists from across the world working there. A real scientific community. He was keen to point out that British scientists had been part of this community, including Sir Edward Appleton, who had published some 'academic papers'.

Cocky had never encountered an academic paper.

Eric was trying to explain the importance of the laboratory they were heading for. At its height, during six years of war, 300 scientists had worked there. Scientists who had previously put their skills to research purposes, such as remote sensing of the cosmos, climate, and Earth's surface. In the run-up to hostilities they had to turn their knowledge to military use, whether they wanted to or not. He explained that Sir Edward Appleton had worked with Dr Dieminger to identify the 'electric blanket' layer[16] that circles the Earth. The main thing about this ionised layer, for radio communication purposes, is that radio waves get reflected back from it. They form 'ground waves' and can travel very long distances along the Earth's curve.

It was hard for Cocky to keep up, but Eric was on a roll.

'Appleton suspected that the Nazis would use this laboratory for some kind of military purpose, so we kept a close eye on it. They used the work of this lab to find new ways of disrupting our British military radio communications. Because Appleton kept in touch with Dr Dieminger, our government knew all about it.

'There's more,' said Eric. 'I flew over this site a couple of times in reconnaissance aircraft. From my observations, it was being used to send "sounding rockets" through the ionosphere, higher than ever before. Beyond Earth's gravity.'

But Cocky was lost in jealousy about Eric's opportunity to fly planes during the war. Eric was six years older, and had flown throughout the war, often making dangerous sorties into enemy territory, being shot at from below and above while operating photographic equipment fitted in the planes. Equipment designed to record aerial images of enemy defences and installations.

Despite his crippling jealousy, Cocky was cottoning on to a sense of being about to discover more than he had bargained for. He was picking up that this mission might be inherently interesting – important, even – and this helped him feel less resentful about the discomfort of the worn-out truck. He had signed the Official Secrets Act[17] just to start this job. Although some people assured him it was just a formality, others made every effort to impress on him the importance of being very careful about verbal and written communication, making sure that he protected state secrets and maintained silence around matters of national security. He had no idea what 'secrets' he could possibly be party to, until he realised that Eric was using this journey to explain some of them. He was being thrown in at the deep end, going to the heart of a German wartime research base.

∗ ∗ ∗

The following conversation includes topics that were covered over the entirety of the mission, over a few weeks, rather than all in one journey. My father says that when he met Eric, before he had even completed his basic RAF training, he remembers himself as a 'know-all' but that Eric really did have the experience and knew his stuff. He concedes that Eric did well to contain his temper for the duration of the journey.

Eric was driving automatically, his eyes on the road but his mind on the task of inducting his new recruit. He was trying to cram a war's worth of expert knowledge into one journey.

'During the war, our job was to stay one step ahead of the Nazi know-how about how they were getting their bombs on target. We kept intercepting messages that gave us information about supplies being taken to places like this.'

Cocky was silent.

'It was Roy Piggott who told me about this particular one. There had been a shipment of what must be casings, with a code name that meant "spicy sausages", but really, if they were getting that tonnage of sausages they would have needed a fridge the size of a hangar.' He chuckled at his own joke, and Cocky wanted to know what 'casings'[18] were, but Eric was off again. He explained that throughout the war, the task of signals intelligence was to stay one step ahead of whatever the enemy were planning. A prominent instance of when the signals[19] weren't deciphered in time was during the later stages of the London Blitz in late 1944 and early 1945. The number of devastating hits proved that the Germans must have worked out how to get their V-1[20] flying bombs and the larger, silent V-2 rockets to hit long-distance targets with accuracy.

'I was in London in 1941 when they bombed the Queen's Hall,' said Cocky.

Eric was relieved that Cocky was finally joining in.

Cocky told Eric about being at the Queen's Hall on the day of the bombing, hanging around outside to get a cheap ticket from box office returns. He had managed to get admitted to an Elgar concert, performed by the London Philharmonic. He had asked staff if he could leave his ARP bike in the lobby, which the staff weren't too keen about. They allowed him to leave it there if he promised to move it promptly at the end, which he did. The next morning, on his ARP bicycle messenger rounds, he saw that an incendiary bomb had destroyed the Queen's Hall, along with the British Museum and big chunks of the House of Commons and Westminster Abbey.

Eric's hands on the steering wheel gripped tightly and his voice tensed. He explained that by the later stages of the war, German bombs had stopped being dropped by airplanes with pilots,[21] as with the Queen's Hall hit. The Germans had learned how to avoid risking the lives of their bomber pilots by developing new, fearsome weapons. Later targets hit in 1944–5, such as Speakers' Corner, Selfridges and Woolworths in London, showed that specific places were being bombed remotely, and with precision. The British had to find out how the Nazis were controlling their bombs.

'Did the scientists working in this lab make direction finders for remote bombs?'

Eric was wrong-footed. He hadn't mentioned radio-controlled direction finders. He knew far better than Cocky about the capabilities of the V-1s[22] and V-2s, as he had been part of the vast effort to understand and intercept them, preventing more of them from damaging the UK. But he always kept the scientific intelligence he gained secure in the correct channels. He would never casually discuss it. To his dismay, Cocky was about to annoy him even more.

'You knew where you were with a doodlebug,'[23] said Cocky, who put his head back against the seat and made the distinctive high-pitched wailing sound of a V-1 flying bomb falling to Earth.

Eric couldn't be sure if Cocky was making some kind of joke. If so, he wasn't laughing.

For Cocky, this was a sound that was far from funny. It haunted his nights, and he associated it with grim scenes of people fleeing in fear and panic, and the aftermath of devastation. His job as an ARP messenger had been to get to the scene of a bomb hit as soon as possible, because the bomb would have taken out nearby telephone or telegraph wires. He would cycle around London with written information on bits of paper. The messages were estimates of the damage caused, for the Air Ministry, War Office, and the BBC. He would pedal, barely able to see or breathe because of dense smoke in his eyes and throat, along streets that suddenly had craters and floods from burst water mains, avoiding desperate people who were searching for relatives.

Eric had to ignore the younger man's lack of tact about the doodlebug. He took some deep breaths. There was another long silence.

'Did "The Doc" come up with any exciting technological advances during the war?' Cocky piped up eventually.

'It's really not done to put it like that,' said Eric sharply, banging his hands against the steering wheel. 'As if "exciting" is what they were aiming for. This laboratory isn't a bloody funfair. You could really offend people by trivialising it. The Nazis wanted their scientists to develop the deadly weapons that would stop us in our tracks. Damn it, Cocky, you are going to offend people right left and centre.'

Cocky was silent, ashamed.

Eric went on, trying and failing to keep the anger out of his voice. 'The Nazis were trying to obliterate the Allied forces. Their Luftwaffe dropped bombs on London, on armament factories in towns all over the UK, and in the final few months of their bombing campaign, when they developed the rocket bombs, the V-2s, they were more accurate and more deadly than anything we'd seen before.'

Cocky was cowed by Eric's anger. He had to make himself speak up so that Eric wouldn't get the idea he thought that bombing was a game. He told him about the bombing of the Filton Works of Bristol Aeroplane Company, when he was 15. About 150 workers had been killed, though no one from his own family, despite four family members working there at the time. It had been a terrible blow for the community and the company.

It was Eric's turn to listen.

'That was the day I decided to do all I could to help win the war. I wanted to get revenge by flying a warplane over Germany. Drop bombs on *their* factories.'

'You wanted to join the RAF from being 15?' asked Eric.

Cocky nodded.

'While you were messing with bikes, I was trying to find out where and how and who was making the bombs, and trying to destroy that capability before towns full of people could be targeted,'[24] said Eric. 'And that's what we are still trying to do, except now the war with Germany is over, the main threat is coming from Russia.'[25]

Cocky had no idea that the Russians were considered a threat.

The two men lapsed into silence as the snow turned to sleet and the rudimentary windscreen wiper tried valiantly to maintain a small patch of cleared glass for Eric to see the truck in front. In the late afternoon, the convoy passed through a series of small towns without stopping. Many towns had bomb damage and others were intact. There was very little for sale in any of the shops and the people trudged around in old clothes, queuing and walking, clutching packages or children's hands, wrapped up against the cold but still freezing in the winter air.[26]

'We're planning to stop in Salzburg,' said Eric. 'But that won't be until morning. We hope to pick up more supplies there. The cities are getting more food through than the smaller towns.'

'The Russians *were* on our side, weren't they?' asked Cocky eventually. 'One of the Allied forces. They fought the Nazis on the Eastern Front, didn't they?'

'The Russians now, and probably even then, are firmly on the side of the Russians,' replied Eric. 'We've known that for a while.'

Cocky tried to digest this change in the political landscape.

'And the Russians are keen as mustard to get their hands on the equipment in this lab,' Eric continued. 'We're trying to get there before they do. That's why we're driving through the night.'

Eric pondered over telling Cocky any more. He was too green and too gobby to trust completely. He might have signed the Official Secrets Act, but he probably wouldn't recognise a secret when he saw one. Eric ventured cautiously on.

'We think, we don't know for sure, though, that you're right about this lab. We think it's where the V-2's direction-finding apparatus was developed. We know the technology used radio waves because we were able to detect them, and we pinpointed the source of them to this lab.'

'We're aiming, then, to get the full picture,' said Cocky.

'Partly. There's another, more personal reason, though, for this mission. I've asked R.V. Jones to set me up with the funding and equipment for my own Air Scientific Research Unit. I want to continue the work we started when it was all about second-guessing the Germans. This mission has to work. You have to help.'

'I'll do my best,' said Cocky. 'With the drawings, I mean.'

'I've persuaded R.V.,' said Eric, 'that unless we keep some British capacity to work out what the Russians are doing, then the Americans are going to overtake us, and they could stop sharing intelligence, which would leave Britain sidelined. With my own Unit, Britain can stay in the game.'

Cocky couldn't keep up. 'What are you expecting to find at the lab we're going to?' he tried.

Eric thought for a while before answering. 'The way I see it, we've got to get to this lab, and strip it of anything we can find, to stop it getting into Russian hands. We don't want them to copy the technology designed for the V-2 and use it for their own purposes.'

Cocky said nothing.

'The official line is,' stated Eric clearly, 'we are relocating Dr Dieminger's laboratory to the British zone of Germany. That's all you need to know.'

'Alright,' Cocky managed, his throat dry.

Eric then tried to lighten things up. He told Cocky how unpopular he was at the moment with 'high-ups' in the Air Ministry, because he had requisitioned all seventy of the available Army trucks. This impressed Cocky most of all, that Eric, despite being an experienced RAF officer, wasn't totally accepted by the men in shabby suits.

'I'm a bit of an outsider,' continued Eric. 'I don't always do things by the book. But this mission is my chance to show them that I can protect British interests. Not that anyone's watching anymore, with the possible exception of R.V.'

Cocky thought it was exciting that Eric might be considered maverick, even if by only one person at the Air Ministry. 'At the Ministry … they are all very … educated,' he said tentatively.

'They've all got lists of university degrees as long as your arm and letters after their names that only they understand,' spat Eric.

'Everyone wears glasses, and the same type of suit,' added Cocky irreverently, forming a pair of thick glasses with his thumbs and forefingers.

Eric tried not to laugh, but a rumble escaped. 'Not everyone has a clue about signals intelligence. Even those with first-class degrees. This war has sifted out the ones that can regurgitate lecture notes in exams, and those who actually understand and can apply scientific principles. I'm afraid that some of the bigwigs are just bloody pretending.'

'Not R.V. Jones?'

'Definitely not R.V. Jones,' confirmed Eric.

'Did you go to university?' he asked Cocky, aware that he could be better qualified than he was himself. It was a loaded question. Eric hadn't gone to university, but to his irritation he was surrounded by people with university degrees, people who were automatically preferred, despite the fact that Eric had years of practical experience with not just flying warplanes, but also with signals intelligence, its methods, equipment, and the science around it.

'No,' said Cocky. He was concerned that Eric might assess his intelligence as low because he hadn't got a university degree.

Eric realised that Cocky was staying silent, determined not to risk his judgments about the route he had taken towards his career in the RAF. He waited, stubbornly refusing to ask for more information, whilst really wanting to know.

Cocky refused to part with any more details about the poverty of his education, unless directly asked. The engine noise rumbled on, the cab lurching rhythmically, the rudimentary wipers keeping snow from totally obscuring the driver's vision.

'Where did you learn your technical drawing skills, then?' asked Eric, in a more genuine, friendly tone. 'How come you're a draftsman?'

Cocky had no alternative but to tell Eric about his education, which was inextricably bound up with his dysfunctional home life – something that he usually kept quiet about. He explained that he had learned draftsmanship, or technical drawing, firstly at school, and then at night school.

'We were always moving schools. Me and my brothers and one sister – we all won scholarships but we would move to another area before we could take them up. So I did night classes. I left school at 14, got a job at Filton Works in Bristol, then did National Certificates in the evenings. Mechanical engineering, physics, draftsmanship, chemistry, chemical engineering …'

'In the evenings, after work?' queried Eric.

'Until the college was forced to close because of the war,' mumbled Cocky.

'Speak up,' said Eric.

Cocky tried to speak up. He explained that the college had closed because of the war, but that some people had managed to do the courses and pass the exams in seven years. He had completed the courses, but closure of the college meant he hadn't taken all the exams.

Eric weighed all of this up. He had experienced a similar route to education, carving out opportunities and putting in the legwork to achieve vocational versions of the qualifications that privileged people took for granted.[27] Yet, this young man had just told him that a lot of people were making up for their educational deficiencies, and studying as adults at night school.

'We, and they, were making up for deficiencies in the educational system,' Eric explained to Cocky. 'Not our own deficiencies.'

'That's true,' agreed Cocky.

'The RAF needs people like you,' continued Eric. 'Resourceful, intelligent, think for yourself, expect nothing to be provided on a plate. Practically taught yourself engineering, and technical drawing.'

'No, I didn't teach myself. We had brilliant tutors. Like Mr Wright,' said Cocky. 'He lent me books and papers, gave me decent pens – I've got one of the pens he gave me here – and we would have a chat afterwards, you know, further ideas about electromagnetism …'

Eric then turned to underlining for Cocky the enthusiastic regard he had for his boss, who was indeed one of those first-class degree academics in a shabby suit and glasses, but who had earned Eric's respect because he was also a 'man of action'.

'Something R.V. Jones did, back in 1942, he masterminded the raid on Bruneval.[28] You must have heard something about it. It was a night-time raid to capture German radar equipment.'

'That was in the papers,' said Cocky. 'The *Daily Sketch*. I remember reading about it. The British parachuted in, managing to keep under the enemy radar, and smashed the Nazi radiolocation post.'

'Well, they didn't smash it,' replied Eric. 'That's typical tabloid twaddle. The whole point of the raid was that we needed the parts intact. They parachuted into Bruneval, *very carefully dismantled* the radar equipment, and carried it to the beach, where they were picked up by a Navy boat off the coast of Le Havre, and returned to Britain with not just the radar parts, but a German radio technician as well.'

'Wow. The *Daily Sketch* didn't report the half of it,' said Cocky.

'It wasn't fully reported, Cocky. There was so much that couldn't be reported back then, especially anything about radar or radio communications. Only a handful of people at the Air Ministry really had the understanding about its role in the war, and the people at Bletchley, of course.'

'Was R.V. one of the few people?'

'Yes he was. And Churchill got it. He saw the value of knowing what the enemy is planning to do, by intercepting their communications. It took a few months of lobbying, but eventually our government prioritised signals intelligence. Churchill's idea was to go to the universities and handpick the best physicists for the work, including R.V.'

Eric had to concentrate on passing a vehicle trying to come the other way, with too little room on the road for the unfortunate car to pass easily. Once back on a clear road, he continued.

'R.V. isn't just an academic. He is that rare combination of able physicist with man of action.' Eric didn't want Cocky to miss his evident respect for his boss.

'What happened to the German radio technician from the Bruneval raid?' asked Cocky.

'He works for me now,' said Eric.

'But the war's over. Didn't he want to go home?'

'He works for me, by choice,' said Eric. 'And he is sort of home. He's a valued member of my Signals Unit in Obernkirchen, Germany. He'll be part of the ASRU when I get the go-ahead from R.V. He's a great addition to the *Kegel* team, too.'

Cocky was feeling sleepy. 'ASRU?' he asked.

'Air Scientific Research Unit,' said Eric.

'*Kegel*?'

'Skittles. Ever played?'

'No,' said Cocky.

'Do you like jazz?' asked Eric.

'Jazz, yes, and classical,' came the answer, with no more concern about saying the wrong thing.

'We've brought along a player and some records,' said Eric. 'Nothing classical, I'm afraid, but plenty of jazz, and as we're going to be holed up in deepest, iciest Austria for a few weeks, we've managed to smuggle along a few lorry loads of British beer! I want people to relax in the evenings. See if we can get some decent food from the villages. We might find a pig to roast. We'll be working our socks off all day. Anyway, it's my trademark as an officer – ask any of them. Bloody good mess parties!'

But Cocky was asleep, lulled to slumber by the jolting, overwarm cab heated directly from the engine, and his relief at Eric's swing towards acceptance and even friendship. The blankets had slipped and he lurched towards Eric on the next jolt. Eric pushed him forcefully back with the arm that wasn't steering, but Cocky still didn't wake up. He slept through an entire early morning stop in Salzburg,[29] where a few officers got out and negotiated to buy up supplies at greatly inflated prices because there was so little food to spare. Local people swarmed around the convoy, trying to sell things they had made, and asking for jobs.

An hour later, the truck stopped again, just outside Ried im Innkreis, at the remote Fraunhofer Institute of Ionospheric Research. The long, low complex of buildings stretched back into the morning mist. Both men in truck 38 and many other drivers along the whole line of vehicles hopped out of their cabs with just one thought in mind. Each found a spot to urinate a little way into the hedges and gateways along the narrow country road. The gentle hiss of many men peeing was, for a while, the only sound other than a little feeble birdsong that could be heard on the still, icy air.

International Gathering of Scientists

The following scenes are creative reconstructions, based on clues and deduction, from the mission to the Fraunhofer Institute. Some specifics are invented, such as 'Wappo' and 'Freddie' as they measure a room, included to illustrate the way that my father couldn't stop himself from 'putting people right' and displaying his knowledge.

At this time, Sergei Korolev was just one of a group of Russians, with a humble demeanour that belied his later role as leader of the USSR space programme. It wasn't until after his death in 1966 that Korolev's name became widely known, even in Russia.

Most people on the British mission were simply drivers, and practical tasks were found to occupy them for the few weeks. A big task was securing enough food for everyone, and the kitchen garden, which even had a gardener, provided fresh winter vegetables. Local people were taken on as casual workers, and this brought pigs and chickens into the economy.

Cocky had his head in a cupboard and was busy sorting through crates of discarded metal and wooden shapes, some with holes drilled through them or wires bristling out of them, or blobs of plasticine or glue still stuck to them. It was hard to see whether they had any value, or to work out what they could relate to. He couldn't decide whether to draw them, sift out the more promising ones, or throw them away.

During the previous weeks, he had followed Eric and Roy around the extensive, rambling laboratories, making sketches and diagrams of anything that they pointed out. He asked them to stick a paper sign on each piece of equipment they wanted in-situ drawings for. He had his work cut out making sure of the accuracy of his section diagrams. It was slow, because he needed to understand the workings of the equipment before he could hope to label everything properly, and he worked on the principle that if he thoroughly understood how something worked, then his drawings would enable others to understand too. Some people were drawing diagrams of whole building layouts, making careful room plans, which they painstakingly paced out, writing down the number of paces it took them to cross the room.

'You're taller than me, Wappo,' said Freddie. 'Your paces are that bit longer than mine. Why don't either me or you take the measurements, then record whose paces we've actually used on the diagram?'

'So, a "ten paces Wappo", or a "thirteen paces Freddie" measurement?' said Wappo. 'That should work. We could use "W" for a Wappo pace and "F" for a Freddie pace.'

'It'll work for exactly as long as everyone involved remembers who Wappo and Freddie were,' said Cocky, without taking his head out of the cupboard. 'And as long as everyone accurately remembers that, in fact, in February 1946, Wappo was quite a lot taller than Freddie. Longer paces.'

Wappo and Freddie peered towards the cupboard. Cocky was rummaging.

'Ah, here it is,' said Cocky, coming out of the cupboard and facing them, a flat, marked metre rule in his hand. 'This is a metre rule. Do you know how long it is? No? Well, it's one metre long. It's always been a metre long, and it always will be. It might expand a little in warm weather, because it is made of wood, but it is accurate to the required tolerance for measuring rooms. Each metre is 100 centimetres long. Each centimetre has 10 millimetres. You could use the internationally recognised abbreviations 'm' for metre, 'cm' for centimetre, and 'mm' for millimetre.'

Wappo took the ruler without a word of thanks.

'Everyone will therefore always be able to accurately interpret any plan that is drawn in metres, and the internationally accepted subdivisions of metres,' said Cocky.

Cocky returned to the interior of the cupboard. Freddie and Wappo started to measure the room in silence, marking down the dimensions on their diagram.

'Cocky git,' said Wappo under his breath.

'Cocky by name, cocky by nature,' agreed Freddie.

'Ah, there you all are,' said Eric, coming into the room at speed, his face red. 'Everyone in the canteen, now. Emergency briefing. Chop chop.'

In the canteen were already assembled Dr Dieminger[30] and his remaining German staff, who were being genuinely helpful to the British party because they were relieved about their imminent transfer to a new laboratory in Lindau,[31] near Göttingen, and being allowed to continue their peacetime scientific studies.

'We have visitors,' said Dr Dieminger. 'The Russians are here.[32] We've been expecting them for quite a while. In fact, at first we thought you were they, when you all arrived – I mean, we were glad, very glad, that you weren't them, but they are here now, and we have no reason not to invite them in.'

'We have no reason not to invite them in,' said Roy Piggott, jumping to his feet in impatience at Dr Dieminger's slow English. Roy went on, speaking fast. 'We are officially still Allies. But this briefing is to tell you, in no uncertain terms, we have *every* reason not to pass on knowledge to them, nor to part with any instruments or equipment we've already loaded into the trucks.'

'We don't show them inside the trucks, we don't show them the "*Apparat*" we have placed in the trucks,' said Dr Dieminger slowly.

'We don't share any of the scientific developments we've seen so far here,' said Eric. 'We can talk to them, of course. We have to be friendly. We are still officially Allies. But we can't share sensitive information that might give them insight into this laboratory.'

Most people had no idea what developments their bosses meant, so there was not much chance of knowledge being passed on to the mysterious visitors. Some people looked puzzled; most of what they had packed up resembled school physics laboratory equipment. There were all shapes and sizes of metal boxes with dials and wires, many forms of electrical circuits, transponders, transducers, miniature wire antennas, metal containers and glass chambers. Some of the crucial papers and notes about the experiments had already been taken by departing scientists.

Wappo and Freddie looked bemused. They were wondering if the dimensions of the room they had just measured so carefully might count as sensitive information.

Four battered Russian long-base army trucks[33] pulled up outside the canteen windows. The vehicles were practically touching the rickety fence that contained the kitchen garden, with its precious plantings of winter vegetables. The windows were steamed up on the inside with so many people gathered together there, so in the few minutes they had available before the visitors found the entrance, Eric gave the order: 'Everyone. As you were.'

Cocky set to work taking the pieces of paper off the items of equipment that he hadn't yet drawn. After a few minutes, he could hear Eric and Roy welcoming the visiting Russians, standing in the entrance hall and giving people time to cover their tracks. Then they showed them into the canteen and Cocky presumed they were being offered some stew. Local people being employed by the British were preparing the stew, and it actually had the luxury of chicken meat in it. The much-anticipated stew would now have to stretch further than expected, but Eric was determined to demonstrate hospitality and friendliness to the group of Russians, with their imposing greatcoats, shaggy moustaches and shiny, knee-length boots.

'Doktor Deemonegger?' queried the apparent leader of the Russians, his wide forehead covered with a fur hat. 'I am Korolev, Sergei Korolev.[34] You know me by repute, no?'

Eric hadn't heard the name, but he was sure that Roy would know, or he could ask R.V. next time they were telephoning. Roy had gone to find Dr Dieminger in another part of the complex.

'More soup?' asked another of the Russians in his own language – a tall, skinny young man with acne and a hooked nose.

'Mitya, you are like my son, always hungry. The son I never had,' said Korolev, in soft Russian.

Mitya smiled back shyly at the unexpected familiarity.

'Of course,' said Roy, who understood a little Russian, signalling for one of the local cooks to come over with more chicken stew.

'You have many British Army lorries,' said an older Russian, who had put away two bowls of stew already. 'Perhaps eighty?'

'Seventy,' said Eric. 'And five motorbikes.'

The Russians waited in silence for Eric's explanation.

Eric kept to the official line. The mission was to relocate a civilian research establishment out of the American zone and into the British zone of Germany.

'The Fraunhofer Institute is a highly respected civilian research establishment,' said Roy. 'The work here is mainly concerned with the ionosphere. You know? The layer of electrically charged particles that forms a barrier between the Earth's atmosphere and the vacuum of space. We believe that the ionosphere protects the Earth from shortwave radiation emanating from the sun. It reflects the radiation back into outer space.'

For ten minutes, Roy[35] kept up a creditable monologue on the topic of the ionosphere before the lead Russian took off his furry hat and held it still, between tense hands, on the canteen table. 'We are scientists,' he said wearily. 'I am Chief Engineer Korolev of Institute Nordhausen.'

The British and German scientists stayed silent.

Roy finally managed to give Korolev the respect he was angling for, even though his name was not known to the British. 'Oh, you are Chief Engineer Korolev? We are honoured to make your acquaintance.'

'My name is Mitya,' said Mitya, in English.

'Olaf,' said a much older man.

The rest of the Russian group stayed quiet, watching and waiting.

'I am Roy, sometimes called Robbo, or Piggott or Piggy. Ha ha – so much for nicknames!' burbled Roy nervously, deliberately fluffing their names so they were harder to memorise. 'This is Acky, and there are other top Brits who you will no doubt get to know.'

'How long do you plan to stay?' asked Eric, who was hoping it would be a few hours at most.

The Russians looked surprised. 'We will stay until our work is done,' they said. 'Now, where to sleep?'

Eric got up wearily to show them their bunks. In the bunkroom, there were hardly any blankets left. He was starting to lift some from other bunks, when Olaf stopped him, went back to his truck, and came out with an armful of coarse, thick, woven Russian blankets in bright colours.

'I'm putting you all together in here,' said Eric, indicating a side room. 'Then you can speak Russian'.

'We are happy to be together with the British,' said Korolev. 'Our wartime allies. We like to talk science.'

'Oh yes, we like to talk science too,' said Eric cheerfully. His face was a picture of bonhomie until he got around the corner and was certain he wouldn't be overheard and couldn't be seen. 'Let's have a look at the winter vegetables,' he said to Roy. They stepped out purposefully into the kitchen garden. 'We have to warn the troops.'

'We have warned them,' replied Roy. 'They know not to give away any sensitive information.'

Eric and Roy stood by the onion bed, looking as if they were discussing which ones were ready to pull up.

'It's more than that, though, isn't it?' continued Roy. 'Korolev, he's some kind of chief, and he must know far more about this place than he's letting on.'

Eric was concerned. 'They might try milking you or me for information. They know we are here for the same reason they are. They'll probably target Dr Dieminger. He's been the gaffer here for years, and he knows the most.'

'We have to talk to Dr Dieminger on his own, make sure he's aware of just who is here, and why they've come all the way from Russia,' said Eric.

'They've only come from Bleicherode, near Nordhausen in the Russian zone,' said Roy. 'Not from Russia mainland.'

'Blimey, and we gave them all that stew,' said Eric. 'I thought they'd had a long journey.'

'Dr Dieminger was piling some more equipment, or "*Apparat*", as he calls it, into our trucks while we had them in the canteen,' said Roy. 'It was worth a bowl of precious stew. And, in case they've clocked which one he put it all into, I've had the trucks moved round a little.'

'The Doc was using the opportunity to save a load of files as well. He was getting people to help him carry in some crates of British beer, so it looked like that was what he was intent on doing.'

'I don't think they're that easily fooled,' said Eric. 'We shouldn't think they don't know we are hiding as much as possible from them. It must suit them to play along with us at the moment, but we mustn't relax, and we mustn't let Dr Dieminger start trusting them either.'

'He doesn't, I'm sure of it,' said Roy.

'It's only a matter of time until the Russians get the idea of doing an Operation Paperclip of their own, and our Dr Dieminger could be first on their kidnap list,' said Eric.

'I've heard from R.V. that's exactly what he thinks they will be planning,' said Roy.

'We'd better keep an eye on The Doc. We don't want him to disappear in the dead of night into a Soviet Army truck,' said Eric.

'Those peas are doing well for late winter,' said Roy loudly. 'Or is it early spring?'

'They're nowhere near ready though; let's see if there are carrots,' said Eric with enthusiasm, even louder.

'Do you have in this garden, cabbages?' asked Mitya, appearing round the corner.

'Cabbages? Yes, over there by that wall. Pull a few up Mitya; they will go well with tonight's stew.'

The task of recording the work of the laboratory was already slow, but with the presence of the Russians, efficiency was impossible. At any time, a Russian scientist might demand an explanation of a piece of equipment, a process hampered by their apparent rule to never give away their level of pre-existing knowledge, and to insist on an explanation that would 'begin at the beginning'.

Eric and Roy went to great lengths not to show interest in pieces of kit that really fascinated them, and to feign interest in things that were widely understood in the West. Everyone had to maintain vigilance about communicating information whilst acting like workmen tasked with the practical job of relocating a laboratory.

The number of pieces of equipment that Cocky was tasked with drawing was reduced to a few of the more complex ones. He took these to a poorly lit storeroom to draw, carefully taking them apart. It was both boring and fascinating at the same time, and the work strained his eyes.

He yearned to read the notes that went with the experiments that these instruments had been used for. His drawings were immediately whisked away when he completed them, and only Eric knew where they were hidden.

Sometimes Cocky wanted to make revisions when a new piece of information came his way, but he never saw his detailed drawings again.

Relations with the Russians were cordial and occasionally warm, but Eric did not feel that he could allow his usual tension-releasing parties to happen in the evenings. Instead, he was keen to spend time with the women from the surrounding villages who he had taken on as domestic staff, and he spoke a soft, mellifluous German with them.

Cocky heard that a couple of the women were sisters who had joined the convoy at Salzburg, and were living in – at the far, disused end of the laboratory. They cooked and cleaned and fetched food from the villages, miraculously appearing with a basket of rye bread, a chicken or a bag of sausages, just as the monotony of potatoes and wheat porridge was becoming unbearable.

In a rare moment of reflection, Cocky was idly watching one of the kitchen helpers as she fed two weaners with a bucket of slops. The young pigs had been donated by villagers, who didn't have enough spare food to feed them. They had specified that they wanted to share in the roasted pork. The much-anticipated feast was supposed to benefit the locals and visitors alike.

Cocky wondered if the two pigs had belonged to the girl's family; she certainly liked being out in the yard with them. She wasn't more than 12, and she reminded him of his older sister Marjorie at the same age, all long limbs and a sharp, watchful face. To his surprise, as the girl spotted his face at the window, she didn't turn away. Instead, she stepped closer, held his gaze, then put down the tin bucket she was holding and beckoned. It was unmistakable. She had beckoned, and kept her eyes steadily on his. He stepped out of the back door and she walked away quickly, looking behind herself every so often to make sure he was following. The pigs snuffled happily in the slops as she led Cocky out of the yard, across a scrubby field, and into a belt of trees on the far side of the recently ploughed field. She didn't say a word.

He followed her without concern about being seen, just a burning curiosity about what she might want to show him. She stopped in a clearing that had been trampled down underfoot and had wheel marks and a bare patch in the middle. She stood on the bare patch and gestured to him to stay where he was. Then she pointed to the trees and branches around. There were black burn marks under the foliage on some of the trunks, and Cocky could see that some trees were deformed, bent and leafless, with no hope of ever coming into leaf again. Then, springing from a crouching position, the girl mimed a whoosh upwards with her hands. She shouted, silently, 'BANG!', and acted an explosion with her outstretched arms, and the path of a missile, going up and up, in a parabolic curve, over towards the houses of the village.

She waited a few seconds for him to understand.

'Did you find metal casings? Did you find anything?' asked Cocky.

'No, everything was carried,' said the girl. 'I looked in the woods, my brother too. No metal-pieces.'

She made sure he had understood. Then she turned and set off back to the laboratories. She made straight for the pig pen and picked up her bucket. She didn't acknowledge Cocky at all, not then, nor during any of the time remaining.

Cocky started to use any bits of spare time, moments when he was sure he wasn't being observed, to look for the used missile parts. He wasn't sure exactly what they would look like, but when he eventually found a huge pile of them under sheets in a ramshackle outbuilding, he was in no doubt as to what they were. His first thought was to get Eric to come and have a look.

It took a while before the two men could be sure of not being followed or seen, and they felt it necessary to set up a diversion in the lab on the other side of the building that would serve to absorb the Russians for long enough for them to examine the pieces of missile. The diversion was a small, complete gyroscope[36] that Cocky had found a few days before, but now he planted it where it would be re-found, and when it was, he and Eric scuttled out of the room.

Their first observation was that these experimental missiles[37] were very small, no bigger than 3 feet long. It didn't seem possible that they would make a bang big enough to be heard from the nearby houses. Each one had been ripped apart by the explosion that had ended its career, and the bits of wreckage mixed with others. It took time for the two men to bag them up and stow them in the trucks, roughly one set of missile parts to a sack. Their fuel canisters were intact, if dislodged, and still had the markings on them to show that each missile had used both ethanol and liquid oxygen.

'There are some of these steel canisters in the fridge in the kitchen,' said Eric, indicating the ethanol[38] ones. 'I thought they were best Russian hooch.'

Cocky looked askance at him.

'Just kidding,' said Eric. 'The Russians have brought lots of seriously lethal vodka, and it's all stacked in the fridge, as you know, but these canisters shouldn't be in a domestic fridge.'

'It's not actually a domestic fridge; that's just the way we are using it,' said Cocky. 'I think it's supposed to be the safe place that the lab had to store the ethanol. There might be a supply of liquid oxygen somewhere too. That'll be sitting upright, in flasks.'

'If we run out of ethanol for the missiles, I'm sure the vodka would do the trick,' said Eric.

'A couple of them were spooning up frozen vodka[39] with their breakfast rolls,' observed Cocky, 'pretending it was jam.'

'Vodka jam,' laughed Eric briefly, before his expression changed completely. 'Seriously, Cocky, if you are offered vodka at any point by our Rusky friends, accept it of course, but don't drink it, whatever you do. It's a well-known trick they use to get people to talk.'

'You are talking to them all the time,' observed Cocky, with some resentment. 'You're not holding back. I heard you explaining how to wire up a lateral accelerometer[40] yesterday!'

'I'm talking School Certificate physics. I'm talking well-worn knowledge about transponders and transmitters, I'm saying nothing new,' said Eric.

Cocky looked unconvinced. The Russians always seemed to be taking it all in when Eric explained things.

Eric hissed at Cocky. 'The lateral acceleration sensing system I was talking about to the Russians, for instance, we all know that it would need to be attached to some kind of analogue computer to be able to measure the deviation. It's useless knowledge without that.'

Cocky felt that Eric was underestimating the Russians' ability to work out that an analogue computer must be involved. He hissed back: 'They know how to stabilise rocket flight with gyroscopes. They could easily make the leap to guidance, and then another leap to the coding part!'

Cocky was talking too loudly, and Eric gestured to him to keep it down. They both looked around them in case they were being overheard.

'I haven't told them anything they don't already know,' whispered Eric. 'It's all in the public sphere, subject to academic papers, albeit not translated into Russian as yet. Nor has Roy or Dr Dieminger. We are all leaving here in a week, sooner if we can. The Russians are wising up to us and they are getting fed up with us talking but saying nothing.'

Cocky had noticed that the Russians' questioning had become more specific, more in-depth, and much harder to wriggle out of by giving bland or generalised answers.

Eric was intent on issuing his warning. 'Tonight, at the jazz party, they might just move in on someone they've not targeted yet. They've been watching who does what, who says what. They've been cornering the ones they think will spill the beans. Look, I'm serious. Some people think, and it could easily be the case, that you know more than most. You do appear to have a brain like a sponge. You don't hold back on explaining things to people, you're unable to stop yourself. It could be you. Be careful, Cocky. Don't, and I mean, *do not*, drink the hooch.'

Cocky blanched. He was already aware that Olaf must have seen him coming back across the ploughed field with the local girl who had shown him the flattened and burned launch site. He had seen Olaf making his way across the same field, and disappearing into the same break in the trees, and he had kicked himself for not being more circumspect. He knew that the Russian party was there to glean knowledge as much as they were, and that everyone was there to help themselves to scientific equipment. But he had got used to them being around the laboratories and he found it hard to be on high alert the whole time.

Cocky found the canteen teapot and poured some cold tea into his hip flask. Then he diluted it to the colour of the Russians' slightly impure vodka, a weak yellow. He lay on his bunk, trying to rest, but Eric's warning was buzzing around his brain like a fly he couldn't swat. He tried not to think about the coming evening.

* * *

Instead, his mind went back to how he had learned the trick about making cold tea look like spirits – from his mother, who lived in permanent fear of violence from his alcoholic father. One afternoon, he and his brothers were playing hide-and-seek in the rented Bristol flat they lived in at the time, and little Bernard had gone into a cupboard and pulled a coat around him, causing a bottle of whisky in the pocket fall out and smash. Their mother had mopped up the strong-smelling liquid and swept up the glass, while his sister Marjorie comforted Bernard, who was crying. Their mother had found an empty bottle and filled it with cold tea from the teapot, putting it back in the coat pocket, the moment before his father came in. For weeks after that, their mother had even less money to spend on food for the family because she was saving to replace the bottle of whisky. Just as she finally had the money, their father had discovered the whisky was cold tea, and had beaten her up anyway. Young Neville and Bernard had watched the attack, more frightening than any bomb, and resolved in their silent anger to get revenge on him one day.

Cocky couldn't relax. He got up again from his bunk and started to pace. He had mentally filed away any knowledge that he considered cutting-edge, any equipment he had found to do with measurement or calibration. He separated out the knowledge and locked it away in his own head. No one could find it there. Even so, he wasn't looking forward to the party.

'Cocky,' came a friendly voice from another bunk. This made him jump. It was Mitya. 'Are you alright?'

'Yes, I'm alright. I'm just tired. I didn't see you there.'

'Can I have a sip of your alcohol?' asked Mitya, getting off the bunk and gesturing to the hip flask that Cocky was gripping tightly.

'It's empty,' said Cocky, 'sadly.'

'I'll fill it for you,' said Mitya. But he detected an edge to his statement, noticing his insistent outstretched hand.

'With 200 per cent proof vodka? No thanks,' returned Cocky.

Mitya seemed to be thinking of another way to get hold of the flask. It wouldn't take a second for him to realise it wasn't empty.

Cocky had to think fast. 'It's supper time. Let's go.'

Although it wasn't quite supper time, he got away with not letting Mitya find out what was really in the flask. It had been a close shave, and he felt foolish for not checking that the bunkroom was empty before lying down, and for holding the flask in such a way that it looked like he was keeping a secret. He was no good at this game of deceit, he decided, yet he would have to sharpen up his ability to lie convincingly if the Russians did try anything tonight.

Over supper, which was chicken stew again but this time served with the potato dumplings loved by the Germans, Dr Dieminger, Eric and Roy were discussing the phenomenon of ionospheric disturbances in radio communication. Cocky found himself sitting next to them because they had been ahead of him in the canteen queue, and now he wished he had distanced himself in case the Russians deduced from their easy camaraderie that he was on the inside track of the behavior of radio waves. He tried to move away to another table, but Eric shuffled up, so there was clearly room for him.

There was no longer any trace of competitive animosity between the British and German scientists, who had been, until so recently, on opposing sides of the war. They even enjoyed unpacking some of the conundrums that had taxed them during hostilities. Allied radio listening stations were still scattered all over Europe. Both sides had listened for information about enemy movements of troops, ships and planes. Radio operators had tuned in and recorded, decoded, translated and prioritised the information according to perceived importance, and passed it up the chain of command so that high-level decision-makers could use it. Now, with these listening stations shared out into geographical zones, the Allies were taking over those that had been formerly run by the Germans and were discovering in them newer and better-built equipment than they had generally been used to in the UK. This helped to increase the perceived value of German scientists.

* * *

Dr Dieminger had drawn some diagrams onto the paper table covering at one of the earlier mealtimes. He had first drawn round a circular ashtray, to symbolise the Earth, and had drawn a larger, dotted line circle around that, to show the ionosphere, thought to be about 60km up from the Earth's surface.

The paper tablecloth was never changed unless it was really dirty, so the diagrams were still there, and Dr Dieminger was in his usual seat.

Dr Dieminger[41] may not have expounded on this exact topic to people having their meals in the canteen. I've included this creative narrative to show his enthusiasm for communicating scientific phenomena and principles, wherever he happened to be. He was known for doing this, and it underlines his genuine desire to be considered a peaceful scientist.

Long waves, explained Dr Dieminger, were long, wavy ropes of electromagnetism. Electric and magnetic fields travel together through space as waves of electromagnetic radiation, with the changing fields sustaining each other like the fibres of a rope. The Doc's hands twisted and flowed together and apart again, one hand always at right angles to the other, demonstrating the movement.

Some people moved away, taking their stew with them to seek out lighter conversation, and some people shifted closer, depending on their level of interest in yet another of The Doc's impromptu lectures. Cocky stayed put. He enjoyed The Doc's methods of explaining things – so simply, and yet so accurately. He found these dinner table talks soothing; the slowness of them gave him a chance to absorb concepts he had grappled with in his studies.

The Doc had fished out the stub of a pencil from his pocket. He wiped his mouth with his sleeve, and started to draw on the diagram of the Earth with its dotted-line ionosphere.

'Radio waves ... are non-visible light,' Dr Dieminger said to the small group still sitting at the table. He spoke slowly because he struggled with his English, but also because the concept of non-visible light was actually hard to digest. Dr Dieminger was sketching on the paper table covering, his pencil point very blunt, his face a picture of rapt attention and effort to convey these fundamentals.

'They can be transmitted from the Earth. Here is a transmitter on the surface of the Earth ... and they bounce in waves, following the curvature of the Earth, and reaching up to the underside of the ionosphere before coming back down to our receiver antennas.'

The pencil line moved up and back from Earth to ionosphere, from sun to Earth and back. He drew slow, snaky undulations that wound their way all around the globe. They appeared to use the ionosphere as a trampoline.

'Look,' he said. 'The curved shape and slow speed of long waves means they can be picked up in faraway countries. A long wave can stretch for thousands of kilometres, or thousands of your wonderful British miles. We can have radio signals moving from continent to continent. Africa to China, America to Europe ... UK to Russia ...'

'Intercontinental,' said Roy.

'Yes, intercontinental,' affirmed Dr Dieminger, liking the sound of the word. 'But, we do have something getting in the way of our looovely loooong long wave.'

Eric was perturbed that Dr Dieminger seemed to have added in a curved line from above that represented the surface of the sun, and now he was scribbling in short straight lines from the 'sun'.

'All this scribble here is natural radio wave activity emanating from the sun. It gets picked up by our receivers just the same as intentional signals, and it causes interference.'

Some people stood to get a better look at The Doc's diagram, now so covered in scribble that it resembled the efforts of a 2-year old.

The discovery of radio emissions from the flaring sun, above the ionosphere, had just been made public to the scientific community, and although findings were ever more simplified in the retelling, the phrase 'cosmic static'[42] was something that the British scientists had heard. The German and British signals experts had already compared situations they could remember of listening in to radio transmissions, when the opposing side appeared to 'jam'[43] the signals.

'It is similar to jamming. It's a natural jam,' said Dr Dieminger. He had been eating steadily, and he now had second helpings of potato dumplings before he continued his lecture.

'The sun is a great big source of radio waves. Most solar emissions are very, very short wave; they are X-rays, ultraviolet and gamma rays. Most get reflected back into space by the ionosphere. But when the sun emits long waves – such as some radio waves, or infrared – when one of these 'flares' is on the sun's surface, there is a spike. Radio receiving equipment experiences this spike as interference.'

Some people peered out of the smeared windows at the grey sky outside. Perhaps they expected to see spiky flares coming from the sun.

'We could have saved ourselves a lot of bother if we had realised that the sun was the source of some of our fade-outs and loss of signal,' said Roy to Dr Dieminger. 'It explains why the V-2's radio guidance system was sometimes so accurate, and then sometimes it missed its target by miles.'

Dr Dieminger chuckled, but he was keen to move the conversation away from military applications.

'How did you find out that the interference was caused by natural radio waves coming from the sun?' Cocky asked, unable to keep quiet any longer.

Dr Dieminger was quiet for a moment. He never let anyone rush his thought process, or the slow, deliberate turning of his thoughts into English. 'We had been experimenting,' he said. 'We needed to find out if it is possible to send a rocket missile beyond the ionosphere, completely through it, with a radiation detector fixed to it. We think, it is our hypothesis, that the sun is the only possible source of all of this ... scribble ... this background radiation.'

Cocky stopped himself from asking further questions. The room was unnaturally empty and quiet, and his voice would be heard, his interest noted.

Most of the diners had scraped back their chairs and returned their plates to the serving hatch. People were checking the rotas for their afternoon tasks.

Cocky stared at his cold, congealing stew. It didn't look very appetising now. He picked up the salt shaker, the smudge of its contents barely visible. He shook it, and then tried to prise it open at the bottom. A small bugging device was lodged between the bottom section and the container. He tipped it into his palm, but it dropped and skidded across the floor. He watched it slide towards the window, and he would have grabbed it, but it was swept up by one of the domestic helpers who was clearing up after the meal. Cocky looked up in shock. Mitya had seen what had happened and was watching him closely from three tables away. Neither man could move to retrieve the bugging device from the bits of cabbage stalk and general muck in the cleaner's dustpan without giving away their knowledge of what the device actually was.

The domestic helper was busy emptying the dustpan into the slops bucket. She poured some liquid leftover chicken stew on top of it, and left it by the door for the kitchen girl to give to the pigs.

Jazz Party

This section contains another constructed narrative, combining Eric's known habit of creating a relaxed atmosphere in the evenings; and an instance of the Russians cornering Cocky in the kitchen, and Dr Dieminger in a storeroom, simultaneously.

In the tiny, functional kitchen at the laboratory, there was a huge, rudimentary refrigerator taking up two-thirds of the available space. It was full of bottles. Two massive topmost shelves were packed with Russian vodka, and the bottom shelf held steel ethanol canisters, which had a distinctive stamp. On middle shelves, bottles of British beer were crammed in tight, facing outwards. The fridge groaned with the effort of keeping it all cold.

From the mess room came the sound of jazz records playing on a gramophone. British RAF and Army members, locals and workers, came in to help themselves to beer, cracking open the bottle tops on the edge of the big enamel sink. A forlorn plate of hard biscuits sat untouched at the side of the room. People were arguing loudly about which record to play next.

The Russians were circulating with their vodka, trying to encourage the British to try it. Only a few diehards could cope with its extremely high alcohol content. A small stream of people made their way to the sink to surreptitiously tip their vodka away.

Wappo and Freddie were daring each other to swallow ever-bigger sips, by immediately washing it down with beer. Not many joined them because the usual warning was circulating – getting drunk around the Russians was dangerous. The small group of people with the greatest distrust of the Russians' motives were feigning a creditable level of bonhomie.

Cocky had tipped away his vodka and replaced it with some of the cold tea from his flask. He didn't even want a beer; he was too anxious about getting cornered. He hovered at the side of the room, exchanging pleasantries with a few people and trying his best to look relaxed. Olaf lumbered over and was soon asking Eric about British and American jazz bands, to which Eric held forth enthusiastically with all he knew – and quite a lot that he was making up on the spot. As long as he was blathering on about jazz in a good-natured way, Eric knew that Olaf could get no further forward in gaining scientific knowledge from him.

Dr Dieminger was doing an awful job of dancing to the music. His glasses had slipped down his nose and he waved a bottle of British beer around wildly. He wasn't anywhere near as drunk as he looked, but he was giving a convincing impression of someone unconcerned about anything but the happy prospect of leaving here tomorrow and setting up a new laboratory at Lindau.

A few other people were dancing by now and Mitya was being pulled up onto his feet by a couple of young squaddies who were eager to show him some of the moves.

Olaf turned to Cocky and asked him to fetch them some more beers, then turned back to Eric. 'So this is modern jazz? Johnnie Dankwort? How do you say it? Dankwortz?' he asked, apparently interested.

Cocky had no option other than to set off for the kitchen to fetch the beer that Olaf had asked for. He was in a state of high alert to the possibility that this could be a trick. On the way, he approached Wappo and asked him to bring the bottles out of the kitchen.

'Wappo, please can you get me a couple of beers from the fridge?[44] They're for Olaf.'

'Get them yourself,' slurred Wappo. 'I'm not your slave.' He smiled defiantly and stayed exactly where he was.

'Cocky git,' said Freddie.

Cocky picked a moment when lots of people were milling in and out of the kitchen, then darted in and felt around on the middle shelves for the beer. He had to look again; someone had moved the British beers to the lowest shelf, where the ethanol canisters had been, and he had to crouch right down. As he did so, he heard the kitchen door shut and the lock turn.

Korolev and Mitya were standing against the door, ignoring the thumps landing on the door from the frustrated British drinkers outside.

Cocky smiled at them and said, 'I'm just getting these for Olaf,' as he stepped towards the door.

'Olaf doesn't like British beer,' said Korolev, 'or British people who don't share what they know.'

'We've shared our knowledge,' said Cocky. 'Dr Dieminger, and Roy, and Eric, have been talking to all of you, sharing with you as much as we can. Much of the original work here, and all of the written reports, have already been stripped out by the Americans.'

Korolev and Mitya hardened their blocking stance in front of the door, refusing to be fobbed off and preventing Cocky from leaving the kitchen. More knocking came from the other side of the door and Cocky distinctly heard Eric's voice say '*Scheisse!*'

'I'm just the man who does the drawings,' he continued calmly. 'I'm a trained draftsman, and I was brought here just to go around drawing the things that the others point out. I've shown you my sketchbook. Would you like another look? It's in the bunkroom.'

'We would like,' said Korolev. 'We want to understand why Doktor Dieminger is so interested in longwave radio communications.'

'It is fascinating,' said Cocky. 'At lunch, Dr Dieminger drew a diagram to show …'

Korolev reached into his pocket and pulled out Dieminger's diagram on the crumpled section of paper table covering, stained with splashes of chicken stew.

'To show what?' said Mitya. 'It's a load of scribble.'

'It is scribble, because it shows interference. Cosmic interference from the flares of the sun,' gabbled Cocky nervously.

'Dr Dieminger was telling us that the sun's flares emit microwaves, which are very *short* radio waves, emitted at a very high frequency. When you are trying to tune in to a very high frequency (VHF)[45] radio, perhaps because you want to listen to some jazz, sometimes interference stops the signal.'

'Quiet,' barked Korolev. 'We have seen the missile launch site. Across the field. We know that you have seen it too.'

'Yes, I have. The missile launch site is most interesting,' said Cocky.

'And we know that you found the empty rocket casings and have stowed them in one of your trucks,' said Mitya. 'Why would you do that?'

Cocky began speaking slowly, trying to give the impression of full disclosure. He explained that Dr Dieminger had sent up some miniature 'sounding missiles', called RMs, from the flattened place in the woods. He was trying to fire them high enough to reach beyond the ionosphere. They had a gas sample collector as well as a radiation detector attached to them. He wanted to find out the composition of gases in the upper atmosphere, and the nature of the radiation there. The sensing device he fixed to the rockets sent back readings that were off the scale of the measurement equipment, proving beyond doubt that the sun is an emission source of natural radio waves. Dieminger had recalibrated the measurement device, and repeated the experiment a few times. Then he had written it up in a scientific journal, published quite recently.

'I've got a copy of his latest article. Eric showed me it. He's written it up in the journal *Natürliches Wissenschaft*, if you'll just let me out of the kitchen to go and get it,' said Cocky.

Korolev was unconvinced. 'We don't believe that all this big laboratory is for hearing jazz music clearly. Or, to see if the sun causes radio interference,' he declared menacingly. 'We Russians are not stupid.'

'There is a rocket casing we found in the barn that still has a radio instrument with a lightbulb[46] screwed into it,' said Mitya. 'Why does Dr Dieminger want to send this ... lightbulb ... to beyond the ionosphere? You're a radio engineer. You can tell us this.'

'We know it's not a lightbulb,' warned Korolev.

'The lightbulb device, it's a radio valve, isn't it, Cocky? said Mitya. 'A vacuum tube.'

'I just make the drawings,' said Cocky. 'I thought it looked a bit like a lightbulb too.'

'You worked at an aeroplane factory in England,' replied Korolev. 'As a radio engineer. You know about radio valves.'

'I was only 15,' said Cocky, taking in the uncomfortable knowledge that the Russians knew things about him that he had never told them. 'I kept the benches tidy. Made the tea. Polished up the metalwork. Swept the floors.'

Korolev looked exasperated.

'And what is this for?' asked Mitya, pulling the tiny gyroscope out of his pocket.

'I'm intrigued by that too,' said Cocky. 'It's like a toy I used to have as a child, a spinning top. No matter how fast or slow you spin it, it finds its equilibrium. It keeps moving in a sort of perpetual motion.'

'There's no such thing as perpetual motion,' growled Korolev. 'The gyroscope, it's stable when it's spinning and resists being pushed, but it does lose its energy. It does stop.'

'I've done some great drawings of it,' Cocky said brightly.

'What's it for?' repeated Mitya. 'How does it stay upright, no matter what happens around it?'

'Let's ask Dr Dieminger,' said Cocky.

Mitya relaxed slightly. He really wanted to know more about potential uses for the small-sized gyroscope.

Korolev and Mitya exchanged glances that also conveyed their readiness to leave the kitchen. The banging on the door had stopped a while ago, but people were still pressed around the doorway, straining to hear what was going on. Finally out of the kitchen, Cocky looked for Dr Dieminger, but he was evidently holed up with Olaf and another Russian in one of the storerooms on the other side of the complex. Eric was frantically looking for a spare key to the storeroom, but the keys had disappeared.

The jazz records were still playing to a depleted number of revellers, and most of those who were still sober were looking halfheartedly for the storeroom keys, pouring unwanted vodka into the pigswill buckets, or trying through various levels of inebriation to work out which window on the outside of the building could lead to the storeroom where Dr Dieminger was being held.

'Cocky, are you ok? Tell me later,' said Eric, without waiting for Cocky to speak. 'We've got to get The Doc out.'

Cocky stepped over Wappo and Freddie,[47] who were out cold on the floor, arms wrapped around each other and faces pressed together.

'There's another way into that storeroom,' he said. 'If you push on the back of the big filing cabinet in the main office, it opens into a cupboard in that room. It used to have all the experiment files in it, but it's empty now. I thought it looked like a false partition wall, so I pushed it and ...'

Eric was moving purposefully towards the main office.

'Eric,' said Cocky, 'they didn't hurt me, or threaten me – not directly, anyway – and if we appear in the storeroom, especially if we climb out of the cupboard on the other side of the wall, we might be putting Dr Dieminger in danger.'

Eric looked uncertain. 'I need to know The Doc's alright.'

'Let's just listen, then,' said Cocky, gesturing to Eric to stay quiet as they gently opened the filing cabinet door.

Dr Dieminger's voice was holding forth confidently, speaking to Russians as yet unknown. It was clear from the conversation that the electric light wasn't working and they were standing in darkness. 'If we can just step outside, where there is some light, I can draw you a diagram that explains all,' he said.

'We think that longwave radio could be used to direct and even control long-range missiles,' said another Russian, 'across thousands of miles.'

'Oh, I don't know about that. All my experiments have been quite modest,' replied Dr Dieminger. 'I've been using very small rockets that fly very high, above the ionosphere, but they only travel a few hundred metres. You've probably found some of the miniature rocket casings in the barn.'

'Are you trying to get the fuel mixture right?' guessed the Russian scientists. 'Is that why there are bottles of ethanol and liquid oxygen stowed around the place?'

'We have to keep them separate,' said The Doc. 'We must be very careful. Otherwise – BANG!'

The Russians muttered to themselves for a while, arguing with each other about their next course of action. In the filing cupboard, Eric and Cocky shifted to prevent their legs from going numb.

'We Germans have a very funny joke about your vodka,' chuckled The Doc heroically, desperate to break the tension. 'We call it ... rocket fuel!'

The Russians made frustrated sounds and kept on muttering. Dr Dieminger was clearly not drunk enough to let his guard slip and they were getting no further with their questions.

The Russians then decided to play the co-operation card.

'It is your duty to share your knowledge internationally with us, your Allied friends,' stated Olaf. 'Your fellow scientists and explorers of the universe.'

'Indeed, genuine scientific endeavour knows no international boundaries,' agreed The Doc.

'We are all working for the greater good of mankind,' said Olaf menacingly.

Dr Dieminger peered gravely through the darkness from one Russian face to the other. 'I am most sincerely interested in keeping the world a peaceful place.'

The Russians had no more to say.

Eric and Cocky heard the key turn in the storeroom door. They unfurled their bodies and backed silently out of the cupboard, padding lightly out of the room.

By the time that Dr Dieminger emerged, blinking and shaken from his ordeal, Cocky and Eric were sitting in the mess room surrounded by empty bottles and sleeping colleagues, chatting amicably.

The next morning, the villagers were up and about, clearing all the debris from the night before into hessian rubbish bags, including the untouched hard biscuits, and serving up a rushed breakfast made of all the ends of the food supplies from the canteen kitchen: boiled eggs with boiled cabbage, hot black tea, and black German rye bread – all helping to disperse a hundred hangovers.

The young girl had come for the weaners, now grown into fine fat pigs, and she led them back down the lane with a strong piece of string. They trotted behind her a little unsteadily, probably suffering from the effects of the vodka in their swill. Cocky wondered if one of them was now bugged, and was unconsciously collecting vital information on the noisy details of life in a pig pen.

The original plan had been for the villagers to share the fattened pigs with the army men, but the jazz party had taken the place of the spit roast normally favoured by the Germans. The girl was now delighted that the two pigs would be roasted and enjoyed by the villagers alone. Cocky smiled at her as she passed on her brisk walk down the lane, but she kept her gaze straight ahead.

'Thank you,' he shouted after her, but she just kept walking.

The Russians left as soon as they had eaten, with no word of goodbye. They took with them all the remaining vodka from the fridge, and many of the canisters of ethanol, as well as cases of equipment somehow moved from the British trucks and into theirs. Luckily, a few cases of decoy equipment had been left in an unguarded truck just for this purpose. It was hoped that the Russians would get it all back to base before discovering that it was agricultural scrap metal[48] from a disused barn.

Eric found Cocky and steered him into the kitchen garden.

'Look, Cocky, what happened to you yesterday evening won't be the end of it. They know that you and Dr Dieminger were holding back on pretty much everything you know, and as soon as this ridiculous pretense of the Russians being our allies is finally over – and it is wearing very thin at the moment – the gloves will be off. You could be in their sights'.

'Unlike Dr Dieminger, though, I really don't know anything,' replied Cocky.

'You know enough,' said Eric.

'I know something about radio communications. I know what thermionic valves are for, and I've got some idea how gyroscopes help to stabilise flight. I've read about long radio waves being used for ground-based radar, for detecting and locating targets. But I know nothing about signals intelligence. I don't know why they didn't latch on to you for that,' said Cocky.

'They did.'

'Oh?'

'The second day they were here, they had me trapped in the barn for two hours. Standing in fresh pig shit, I was; my boots will never be the same.'

Cocky was stunned. Eric had been unfailingly polite and friendly to the Russians. They had asked Eric to corroborate their knowledge about the locations of the listening stations across Europe, and any special features about them. Eric had told them that all the ones he knew were being amalgamated, relocated or closed down, but was vague about their purposes and personnel. He told them that the Signals Unit in Obernkirchen was being disbanded, which was technically true. Eric would be using the same site for his Air Scientific Research Unit.

Cocky had been waiting to tell Eric about a conversation he had overheard a few nights ago, but with all the drama around the jazz party, he hadn't had the chance. He and Roy had overheard the Russians discussing matters at Institute RABE,[49] in Bleicherode, Nordhausen. They had gained the impression that it had immense size and ambition.

'It's focused on replicating the V-2 rocket, and making lots of them. Hundreds are already working there,' said Cocky. He realised by Eric's lack of reaction that he already knew all this.

'The Doc has several of his former staff members working there,' said Eric. 'We already know a lot about it, through The Doc keeping in touch with them.'

'Our Russian visitors drove over from Bleicherode,' said Eric. 'That's where they're heading back to today. According to R.V., they are some of the top Russian scientists. R.V. told me last night that Korolev was on a list of the scientists Stalin had banished to a labour camp before the war. Stalin banished him to a gulag, for "sabotaging" some weapons work he was involved in, back in 1938. He was only reinstated last year to a senior position. He now appears to be tasked with gathering displaced scientists into one place. No one really knows why.'

Cocky looked shocked. How could Korolev be accused of something as serious as sabotage?

'Sabotage to the Russians can mean anything from working a little too slowly to expressing doubt that an imposed deadline can be met,' explained Eric. '"Sabotage" is not what we would call it.'

'Roy told me something more about it,' said Eric. 'For six years, Korolev was made to work in a gold mine, or a salt mine; Roy's not sure if he said *'goldt'* or *'saltz'* so I don't know what sort of mine it was, but basically, it was because Korolev wasn't getting very far with his project.'

'Working slowly,' said Cocky flippantly. 'Banishable offence.'

'Anyway,' said Eric, 'after six years of banishment and really hard labour – Korolev told Roy he lost his teeth because of malnutrition – Stalin offered restitution to his former post. He got a nice apartment for his wife and child, who had become practically destitute in the meantime. All this was quite recent – he was only released in 1944. When the Kremlin restored him, he was made Chief Engineer at Nordhausen.'

'I've got a theory about what they might be doing at Nordhausen, why they're gathering scientists there,' said Cocky. 'I heard them talking about trains – trains leaving from Nordhausen station, and going to Moscow.'[50]

'How did you know what they were saying?' asked Eric incredulously.

'They drew it,' said Cocky, 'on the back of a Charlie Parker album cover. The one with the "Summertime" track. After they'd gone to bed, I took the cover to the storeroom where I do my drawings, copied their sketch into my book, then put the album back in the pile.'

'That's one of my best album covers!' sputtered Eric. 'The cheek!'

'I've got my copy of it here,' said Cocky, pulling out his technical drawing book.

Eric studied the drawing with interest, all the time aware that someone could be watching them, so he needed to pretend the drawing showed just another piece of equipment.

'Loading scientists onto trains will never work,' said Cocky.

Eric looked cynically at Cocky. 'They won't be nice about it … Oh *do* get on this nice comfortable train, you are going to *love* your special and very long *holiday* in deepest *darkest* Siberia.'

Cocky shifted uncomfortably. He could imagine the Russians employing forceful tactics to get the scientists onto the trains.

'It's not a mad idea. It's a similar idea to Operation Paperclip,'[51] said Eric. 'Except in that case, the scientists wanted to go.'

'Even if European scientists don't mind living in post-war Russia,' replied Cocky, 'they surely don't want to help the Soviets develop weapons that could be used against their own countries?'

'Some Germans might go willingly. After all, they were supporting the Nazi regime until recently, and not all of them can be innocent. Russia would be the perfect hiding place from the international courts,' surmised Eric.

'Blimey,' said Cocky.

Eric continued. 'If the Russians are trying to do the same as the Americans did, selecting the people they want to work with on their own weapons programme, then they will be compiling some kind of hit list.'

Cocky tried to absorb the implications of this. 'They might have your name down, Eric.'

'Another thing,' continued Cocky, 'Roy heard Korolev say that rocket technology can be used to explore space.[52] He was joking around one evening, but what he actually said was, "One sort of rocket points *upwards* into outer space; the other flies *along* to land on a target in a specific country."' Cocky mimed the trajectories of the rockets with his hands, just as Roy had explained it to him.

'I can't see the point of going up into space,' said Eric. 'It's an interesting thing to do, fascinating even, but why would it be a priority for the Soviets? Its economy is struggling as much as Germany's, and Britain's. Just like us, it has to rebuild.'

'Perhaps it was a joke,' replied Cocky. 'Korolev is interested in exploring space, but his colleagues would know that it had been a big mistake for him to mention it once before.'

'R.V. Jones confirms that Korolev was banished for six years,' said Eric. 'Stalin is paranoid and unpredictable, but I can't believe that even he would banish Korolev for being interested in space.'

'Stalin might have been punishing scientists who weren't totally committed to missiles,' said Cocky. 'Perhaps he didn't want to fund dreamers?'

'It's possible, I suppose,' said Eric. 'Space travel has entertained people's imaginations for centuries. Rockets have been around since medieval times. Going beyond visible Earth, beyond gravity, into unknown space …'

Cocky's imagination was taking in the enormity of the idea.

Eric interrupted his thoughts: 'Stalin is only interested in missiles. He probably wants a long-range missile, as fast and silent as the V-2, that he can direct from one continent to another.'

'A long-range radio-controlled missile,' mused Cocky.

'The difference, technically, is that a missile has a bit of it that explodes,' Eric explained. 'A bomb, attached to the space rocket.' Eric ate a handful of green peas straight from the pod.

'They found out that The Doc can't be bribed,' said Cocky. 'Roy and I heard the Russians trying to interest him in a luxurious dacha on the outskirts of Moscow, a car with chauffeur, all the material goods he could ever want, and girls too. You should have seen his face! He turned them down so politely.'

Eric and Cocky laughed briefly at the thought of the strait-laced, married Dr Dieminger being offered girls.

'I heard something really important about Korolev,' said Eric, not wanting to be outdone. 'It was shortly after they all arrived. Korolev and Roy were in the kitchen garden looking at the pigs.'

Cocky looked up, interested.

Eric went on. 'I was in the barn, with the local helpers, sorting through the potatoes. Korolev was asking about slaughtering the pigs, hoping there would be a spit-roast. Roy started asking Korolev how the Kremlin support their state scientists. Korolev emphasised how comfortable his family is now – their apartment and car supplied by the state. He's totally giving the Soviet-approved answer. Not a word more, or less.'

'If you want my opinion,' tried Cocky experimentally.

Eric waited. He hadn't actually asked for Cocky's opinion.

'I think,' said Cocky, 'underneath Korolev's hard surface is someone damaged by the Soviet system, who lives in fear of losing his liberty, his home, his position and even his family, unless he plays along. He's probably extra loyal to the Soviets on the surface – absolutely squeaky-clean – because he's experienced being an outsider. He nearly died from it. He can't go back to that. He can't have his family[53] going back to that either.'

Eric nodded in accord.

Cocky didn't want the sweet moment of Eric's agreement to end. He said nothing more, in case he spoiled it.

Eric considered this summing up of the Nordhausen Institute's chief engineer. 'It might explain why he's menacing, but he's not actually cruel. He's going as far as he can to scare us, but no further. His heart's just not in it.'

'He certainly had me scared, when I was trapped in the kitchen,' said Cocky.

'And Mitya,[54] what about him?' asked Eric.

'Korolev's apprentice,' said Cocky. 'In many ways, scientifically and personally, I think they are really close. Not his son, but he might as well be.'

Eric wiped his hands on his trousers. 'Cocky, I have something to tell you. I've done it to protect you. I was, before all this, wanting you to join my Air Scientific Research Unit. That was my plan before the jazz party.'

All Cocky wanted to do was go to work with Eric at the new ASRU.

Eric explained that it wouldn't be safe to do any of that yet. He had noted the Russians' interest in his Unit, and in Cocky, and his solution was to get him posted to Singapore.

At this point, Cocky thought he had blown his chances, that he had been too indiscreet around the Russians. 'I'm much more careful now than I was at first. Much more aware,' he said defensively.

Eric reminded him that by now, the Russians would know about his engineering qualifications, the Bristol Aeroplane Company where he had worked, and even where his mother lived. Whether this level of personal knowledge could be true or not, Cocky was shocked at the mention of his mother. The last thing he wanted to do was put her in danger.

Eric had already spoken with R.V. Jones, and they had agreed that Cocky would be much safer designing aircraft hangars in Singapore for a couple of years.

'Your ship, the *Dilwara*, is sailing from the UK when you get back,' said Eric. 'Brush up on your draftsmanship. Learn some Russian. Have some fun. I'll be in touch again when you get back.'

Cocky looked up from the sweet winter peas, ready to argue.

Eric grinned; he had something else to say. 'Oh, and by the way, R.V. has agreed to promote you to sergeant. Sir!'

RAF Changi

The following account is taken directly from my father's biographical notes, as it is relatively fully remembered and recounted by him.

Our troopship was the *Dilwara*, an 11,000-ton ship that was built especially narrow so that it could get close inshore with troops. It rolled like a pig. One good thing about the ship was that it had an excellent library on board. Various groups of the different services left the ship at

Suez, Aden and Colombo. After a month at sea we arrived at Singapore and a whole different world.

Japanese POWs awaiting repatriation to Japan were working in the docks. Our journey through Singapore city to Changi RAF camp took us through mainly Chinese quarters. Out of every balcony, long bamboo poles carried rows of black washing. Leaving the city, we passed the padang [playing field] where the local British played cricket, the Raffles Hotel, and Kallang Airport, at that time the civilian airport of the island. Further up the road, we passed the infamous Changi Gaol. It was dark when we arrived at our billets.

I was in charge of the drawing office with two staff, one of whom was a Chinese clerk called Wee Chang Bun. He spoke good English and was intent on improving his reading and writing skills. I used to go through his self-imposed homework exercises and discuss them with him. I'm sure that when Singapore became an independent state, he would have secured high office.

My own work involved drawing aircraft parts that could be made in the workshops at Seletar. Another project was to provide a portable shelter to protect crews who worked in jungle airstrips. As well as designing hangars for two proposed new runways, I also got involved in the groundwork for the runways themselves, which were planned for an area that was then a mangrove swamp. I mentioned to the Groupy that George Stephenson had dealt with a similar problem at Chat Moss, when he built the Manchester–Liverpool railway, and had filled swampy ground with trees. So, all the palm trees that could be got were delivered to Changi and instantly sucked into the morass. Month after month, this went on with no visible result.

My office contained several boxes of Japanese instruments and equipment from their aircraft maintenance depot. I was asked to inspect these items and report on their quality. Some of them were Japanese copies of German instruments, but all were superior in design and manufacture. The Japanese capture of industrial markets after the war was no surprise to anyone who'd seen this stuff.

Leisure time was spent at RAF Changi Yacht Club, where I was a member of a syndicate that had the use of a dinghy. Most Sundays we used to get sandwiches from the cookhouse and a cooler of beer from the yacht club bar, sailing to one of the islands in the Johore Strait and swimming in 80-degree water. On Wednesdays and Saturdays, there was a series of races for a trophy. Our skipper Harry was an army corporal in the Royal Signals. He was a superb sailor and knew every shoal and current on the Strait. His skill made us win the trophy twice running!

I also taught Technical Drawing at the camp education centre, and presented a classical music programme on Radio Changi, as well as being one of the duty announcers there.

Other groups of servicemen came and went, all doing a two-year stint, but my father wasn't sent back until the very end of the RAF's time there, again because his role was hard to replace. He says: 'I was actually the longest delayed because, apparently, there wasn't another draughtsman in the whole air force who could be sent out.'

Chapter 3

Air Scientific Research Unit, Obernkirchen, 1951

Eric and Marianna

'When I said I would see you when you got back from Singapore, I had no idea you'd be out there for four bloody years,' grinned Eric Ackermann, as Cocky stood by Eric's car after arriving at the little railway station at Obernkirchen, near the RAF Bückeburg base.

Cocky had been expecting to walk the short distance to the Air Scientific Research Unit in Obernkirchen, the one that R.V. Jones had facilitated for Eric. It was a nice surprise to be met at the station by Eric.

'I had a moment in my schedule, and I've been wanting to see you again properly,' said Eric. 'Your security clearance has only just come through from the Air Ministry. I've been moving mountains to be sure you'd at least be offered the posting here in Obernkirchen.'

'I was pleased to get it,' said Cocky. 'I've been holed up in the London office for months.'

'All the Air Ministry staff I used to know there have moved on, and the new lot took some persuading to get your papers sorted,' explained Eric. 'In the end, I had to ask R.V. Jones to intervene. He's still in touch even though he's now a professor at Aberdeen University.'

'Oh, R.V., "Harvey",' quipped Cocky. 'I've never met him but it always feels like he's pulling my strings.'

'He's rock solid,' said Eric. 'He really gets what we're doing at the ASRU.'

Cocky was keen to know what went on there. He knew it would be more interesting than designing the roof trusses for aircraft hangars at RAF Changi, or arranging for hundreds of old palm trees to be dumped in a mangrove swamp to create a runway surface. The four years had flown by like a holiday.

Germany was a complete contrast. Even so, there had been no need for Eric to meet him at the railway station.

The real reason became evident as they walked towards Eric's new car, of which he was inordinately proud and was inviting Cocky's admiration for it. Eric's whole demeanor was different now – his pride in the ASRU and in his car was palpable.

Cocky had filled out a bit, his years of relative leisure in Singapore being the first time he had ever put on weight. At 25, he was also self-assured, without lapsing so often into cockiness.

Eric appraised Cocky with evident approval.

'I had no idea I would be out in Changi for that long, either,' said Cocky. 'I had to stay until the very end – a few of us were supposed to do the handover to the locals. Suddenly, the last troopship was sailing, the *Empire Windrush*,[1] and the last of us just scrambled to get aboard.'

Eric walked around his car to open the passenger door for Cocky, who knew full well that some expression of admiration was expected. He held back. He threw his bag on the back seat and settled into the comfortable passenger seat.

'You've certainly grown up a bit. Let me have a look at you. Have you left a girl behind in Changi?' said Eric.

Cocky choked at this unexpectedly personal question, before adjusting his bag in the back of the car. 'Nice car,' he said to Eric. 'What is it?'

'It's a Packard 210. My pride and joy.'

Cocky admired aspects of the interior and dashboard until Eric had forgotten his question about the possible girl in Changi.

Eric had more news. 'The Packard's my *second* pride and joy, anyway. Since I saw you last, I've got married.'

'Married?'

'Don't look so surprised,' said Eric. 'People do get married, you know.'

'Oh, it's just ... I thought you were already married ... to someone in England?'

'Well, yes, I was married, to Dorothy. But we agreed to divorce, and I got married again a couple of years ago, to Marianna. She's actually Hungarian, though everyone assumes she's German. You do sort of know her – do you remember two women we picked up in Salzburg on the way to the Fraunhofer Institute back in 1946? They wanted work, and Piggott offered them jobs as domestics at the Institute? It was supposed to be just for the time we Brits were there, but that's not how it's worked out. Marianna's my wife now.'

Cocky vaguely remembered the two women. The convoy had pulled into Salzburg early in the morning to pick up food supplies, and although he had slept through the whole episode, he knew that local people wanting work had mobbed the trucks. He heard that a couple of young women were taken on – sisters – but he didn't have much recollection of them. He remembered, though, how keen Eric had been to spend time hanging around the canteen and talking to the workers through the kitchen hatch. At that time, Cocky had been too naïve to think more of it. He had assumed that Eric was organising the catering. A few things started to make sense. Cocky remembered that Eric hadn't shared the truck journey with him on the way back. Now, he realised that he had been more interested in sharing his warm cab with Marianna.

'Congratulations,' he said to Eric with a grin.

Eric grinned back. 'We're inviting you over to our house for a meal tonight. Nothing fancy, but it'll give me a chance to fill you in properly with all you've missed. I've got the Air Scientific Research Unit up and running a treat. Fully staffed, fully equipped, it's bloody great. Our house is just down the road from it.'

Cocky could see how settled and happy Eric was now. He wondered, for a brief moment, if he would ever feel the same way.

'You'll have your own office sorted by the end of the week,' continued Eric. 'I wasn't absolutely sure you'd be coming so I didn't presume you'd need it. I'm thinking that a small team of my best signals men will do the job – are you okay with that?'

'You're forgetting that I've never actually done any signals intelligence work. I hadn't done any when we went to the Fraunhofer, and I still haven't done any. The team will know far more than I do. They will have to be the ones to teach me.'

'You'll be fine; you'll pick it up in no time. You've a brain like a sponge, I've always said. And anyway, the Air Ministry have rubber-stamped it.'

Eric couldn't wait to share some of the headline news of the last few years. 'Our friend Korolev – remember him? He's been busy. He's heading up a whole operation to upgrade the V-2. He copied it, and has modified it to reach a range of 600km. Their version, the Russian copy, is called the R-2. Production started last year, and now the Red Army have hundreds of them.'

'Blimey,' said Cocky. 'What are they doing with them?'

'Stockpiling them at the moment. And, we know *exactly* where they are.'

Cocky grunted in amazement.

'The Fraunhofer Institute that we dismantled', explained Eric, 'is now established in Lindau, near Göttingen, and Dr Dieminger – remember The Doc? – is still the director there. He's working out how long-range communications can be improved, by siting powerful transmitters actually in near space.'

'Long-range communications,' repeated Cocky. 'I see.'

'And our own rocket guidance systems haven't stood still, either,' said Eric. 'We're working with the Americans to really develop them. We're using telemetry to analyse the processes that occur in flight, then feeding that back into the control systems. It's the beginnings of a new sort of guidance system, which doesn't get scuppered by interference. I've actually got one at home to show you. It's got three gyroscopes set apart from each other, which track changes in the direction, angle and roll of the missile; and an accelerometer to measure the speed. All the data is wired to an onboard computer. We're stuck on how that translates to the missile changing course, but it's a start.'

Eric's face was full of enthusiasm that he wanted to share, but it was too much to take in.

'Whoa,' said Cocky. 'Slow down. I'm boat-and-train-lagged, and I need some food.'

'After dinner, then,' said Eric, pulling up at his house and waving wildly as Marianna came out of the front door, pulling her apron off as she went.

'I've made goulash,' she said, hugging Eric and smiling at Cocky. 'I hope you like it with a bit of fire!'

The meal was the best that Cocky had tasted since Singapore. Marianna clearly knew everyone at the Unit, because they shared meals on a regular basis, if Eric wanted a confidential chat with someone.

Marianna was attentive to Cocky, picking up on Eric's wholehearted welcome. She clearly adored Eric, tending lovingly to his comfort and seeking reassurance from him about her use of English, or the amount of paprika in the goulash. Cocky found her delightful.

'Goulash is my favourite from Hungary,' said Marianna.

The table was set neatly with silver cutlery, flowers, and a tall, beautiful water jug in the centre.

'Did your family escape from Hungary?' asked Cocky.

Eric shot him a look that begged him not to ask Marianna about her traumatic flight from the Red Army in 1944.

Marianna sent Eric a look that told she wanted to tell her story.

'Just me and my sister, Georgy,' said Marianna. 'I don't know what happened to my parents.'

Eric stood up to serve the goulash, making small talk about how much busier the little station at Obernkirchen was becoming.

Marianna waited for him to stop talking, and then she continued: 'The Russians had taken over Budapest at Christmas 1944. It was Easter by the time the Red Army came to our town. We were very scared, because we were young women. There were terrible, terrible stories about the Russian soldiers. We escaped under the cover of a lorry going to Vienna. But there were too many people all arriving there, so we bartered some jewellery and got another lift. On Good Friday, we ended up in a little farming town on the Austrian border. We worked there until the end of the war.'

'A great day – 7 May 1945,' interjected Eric. 'End of the war in Europe'.

'When the war ended we went to Salzburg looking for work,' continued Marianna. 'I wanted to be a typist, and Georgiana wanted to work in a flower shop. But there was only factory work, or building work, so we did that.'

'Until our convoy came through Salzburg on a winter's day in February '46, on the way to the Fraunhofer Institute at Ried im Innkreis,' prompted Eric. 'Your knight in a shining army truck.'

Marianna smiled at the memory. 'Yes. Your big line of trucks. We all thought you had brought food, and sweets, or cloth for clothes – we were very excited.'

'We did buy food,' said Eric. 'We put money into the town.'

'But there wasn't enough *food* in the town,' countered Marianna. 'You took a lot of food out, which we needed. All the people were hungry.'

'That's why you were arguing with Roy Piggott that day, remember? That's why you and Georgy came with us, with our convoy, to have jobs cooking for us at the Institute for a few weeks.'

'It was blissful,' said Marianna. 'We had chickens from the local farmers; we made chicken stew.'

'I remember the chicken stew,' said Cocky.

'Do you remember me and Georgy?' asked Marianna.

'No, sorry, I don't. I only remember a local girl, aged about 12. She seemed to know the pigs that were kept next to the kitchen garden.'

'Elise? Yes, she was the farmer's daughter,' said Marianna. 'They were her family's pigs.'

'When the Russians arrived at the laboratory, we made sure they did not hear us speak, or notice us in any way. We hid in the kitchen, pretending to be local, not saying a word.'

'That'll be why I don't remember you,' said Cocky. 'You were undercover.'

'And afterwards, you came back here with me to Obernkirchen,' said Eric, catching Marianna's eye and trying to move the conversation on to happier memories.

'Yes, I did,' said Marianna.

'An unexpectedly wonderful outcome of the expedition,' said Eric, helping himself to more rice.

Marianna smiled over at him. 'Wonderful,' she confirmed softly.

Cocky ate hungrily. He had missed much of the immediate fallout of the war in Europe through being in Singapore – all the displaced people, all the heartrending stories of loss. 'You and I are lucky,' he said to Eric eventually. 'Apart from when you were piloting bombers, you were battling with radio waves, not with guns and soldiers, and during the war I was too young for anything more dangerous than ARP duty.'

'The Battle of the Beams,'[2] said Eric. 'Still very much a battle. Still battling bombs.'

Cocky and Marianna exchanged glances. Neither of them wanted to challenge Eric, but it was not clear to either of them, particularly to Marianna, how signals intelligence had made a difference to the outcome of the war. There was no public information about it, just mysterious antennas and listening stations, units and outposts around Europe, with high security around them.

'Have the Soviets become the new danger, like you feared they would be?' tried Cocky. 'Are you keeping track of their activities now?'

When Marianna stood up, Cocky noticed that her dress was stretched over quite a bump.

'We will do the dishes,' said Eric. 'You need to rest.'

'Thank you for the goulash,' said Cocky. 'It was the best food I've had in ages.'

'You're welcome. It's nice to meet you,' said Marianna.

After washing up, Eric and Cocky moved into Eric's study. It was wildly messy, with bits of electrical equipment on the floor and shelves.[3] It was a calming room, dedicated to thinking things through, and Cocky relaxed completely. He was enjoying Marianna and Eric's hospitality, and starting to feel excited about his new job, which, as yet, he knew nothing about.

Eric was impatient to speak. 'Dieminger in 1946 had access to V-2 technology. He was using those small rockets, the RM's, to blast through the ionosphere and get above it, into space, to measure the levels of radiation, temperatures, gas composition. He was sending these rockets up, with instruments attached, and with *cameras*, to work out how artificial satellites in space could be used for photo surveillance. That's how we know *exactly* where the Red Army is storing its R-2s.'

'I remember the diagrams The Doc used to draw,' said Cocky, 'and his lectures.'

'Yes, but listen,' said Eric. 'The main reason the Russians were there wasn't the experiments at all. Not spy satellites, or intercontinental missiles. It wasn't the equipment that you spent ages drawing. Remember when they tried talking The Doc into coming to work for them?'

'I thought that was a vodka-fuelled joke,' said Cocky.

'Not a joke. They offered him a job, a flat in Moscow, a dacha in the countryside ... and unlimited girls. And they meant it.'

Cocky remembered the Russians' amusing offer to Dr Dieminger, but he was stumped.

'They wanted the scien*tists*,' said Eric.

'We suspected they were finding out *who* really understood things,' tried Cocky, 'so they could, what, kidnap them? Take them to Russia? On those trains?'

'They've done it,' said Eric. 'The group of Russians we met, they've done exactly that. The place they came from in Bleicherode, the Nordhausen Institute, the Russians were gathering international scientists there. Headed up by Helmut Gröttrup, director for the German scientists, and Korolev leading the Russians. Gröttrup – he had been Wernher von Braun's[4] deputy at Peenemünde. In October 1946, later that same year, the Russians piled all these scientists and their families onto ninety trains.'

'They were forced to go?' asked Cocky.

'Yes, they were forcibly deported from the Russian zone. Dieminger knew some of them and they told him the KGB did an early morning raid with guns,[5] knocking on the doors of people's homes.'

'Why didn't they run?'

'Not everyone was against going,' said Eric. 'There were no jobs in Germany at that time. This was a solution of sorts. Once they were on the train, the Russians promised Gröttrup that everyone would live well; they just had to work with the Russians for five years.'

'But these scientists are being forced to help the Russians make weapons – weapons they could use against the UK, Europe and America!'

'And what safer place to be when super-powered missiles start flying about, than tucked away, safe in the immense wastes of the Russian countryside,' said Eric. 'Some, as I say, went willingly.'

'I wouldn't have boarded the train,' said Cocky.

Marianna appeared at the door with a tray of Hungarian pastries. They were delicious.

When she had gone, Eric tried to give some context. 'The world was in shock after the atomic bomb that the Americans dropped on Hiroshima in

1945. Russia wanted – still wants – to develop a nuclear bomb. We know where the places are where they are developing nuclear weapons.'

'ASRU is collecting intelligence about these sites?' queried Cocky.

'Yes. We are finding out where the rockets are being designed and made. They've moved on from the V-2 and are already developing longer-range ones. Powerful rockets, coupled with nuclear warheads. That's what we're trying to stop.'

Cocky thought about the enormity of the task. 'Russia is immense,' he said, getting up and going over to a map of the world on Eric's wall. 'It's got mountains and lakes and rivers and mines and an awful lot of countryside. It's fifty, no, seventy times the size of the UK, and twice the size of America.'

'Yes. But, we can already detect the signals from their test sites. We know when a missile test flight is made and how successful it is. You will be helping us build a complete picture of their activities, from your comfortable office at the Air Scientific Research Unit.'

'Who is "us"?' asked Cocky.

'It's not R.V. anymore,' said Eric. 'We've now got Government Communications Headquarters. We fly all of our intelligence over to GCHQ every day. They are giving me a second-in-command, a senior scientific officer, to hurry things up, especially if a nuclear strike is imminent.'

'Imminent?'

'Look, we know it's not imminent,' said Eric.

'But how do we know, how can we be sure they aren't tricking us into thinking it's not imminent?'

'I'll show you tomorrow how we know,' said Eric wearily. 'For now, let's call it a night.'

Signals Intelligence

My father's memory of the Air Scientific Research Unit in 1951 is that it was run almost unbelievably informally. Eric was rarely there, and when he was around, he hardly ever wore uniform, and he didn't care if others did or not. The actual work took place amongst a great deal of good-natured chat, a fact that belied the importance of it – analysing radio signals from Soviet Russia, to gain insight into their efforts to create atomic warheads and fly them great distances on rockets.

Collecting and deciphering military signals was hard, repetitive and frustrating work. In the short time he was there, my father was loosely tasked with inventing or improving any device to make any part of the process easier.

He remembers skittles nights being treated with the same level of seriousness as the work; there were frequent diversions such as homemade plays and musicals in the evenings, and there was a mess room with a bar. Eric would sometimes treat his staff to a blast of *The Goon Show*, relayed through the camp on loudspeakers, as they were nearing the end of an arduous shift.

The only clothing choices Cocky had in his bag for his first day at Obernkirchen were his still pristine RAF uniform, which, annoyingly, his mother had ironed sharp creases into, and a casual linen beach shirt adorned with blue embroidery. Along with baggy khaki shorts, these were all he had needed in Singapore's tropical climate.

Cocky tried to see what other people were wearing from his window. He had one of two beds in the room, but no roommate, as yet. As people walked by, uniforms were occasionally to be seen, casually draped over other clothes, so he decided to go for his beach shirt with his uniform jacket on top. He picked up his technical drawing kit and made his way to the canteen for breakfast.

'New arrival?' asked a friendly red-faced man in front of him in the queue.

'Yes, I just got here yesterday,' replied Cocky.

'I'm Paddy,[6] and this is Dickie.'[7]

'I'm Cocky,' said Cocky, giving his RAF nickname, as he could hardly remember his real name.

'Draftsman?' asked Dickie.

'I'm a scientific officer,' said Cocky proudly.

'Oh, an officer, eh? I'll have to mind my Ps and Qs,' laughed Dickie.

'I'm Geoff.[8] Pleased to meet you,' said another man, with a shock of untamed red hair. 'I'm a draftsman, too. I recognise your drawing kit. British Thornton. Very nice.'

'Present from an old tutor,' said Cocky.

'Have you got a set of trammels in there?' asked Geoff. 'I've lost mine somewhere along the way.'

'Yes, mine are practically unused. All the compasses are present and correct,' said Cocky.

Paddy looked over with interest. 'Ooh, I'll come and find you when you're settled in and have a root around in your box. I need a compass with a sharp point for some parabolic antennas.'

Cocky smiled at Paddy. 'I'm not sure exactly where I'll be. I need to meet up with Eric to find my office and my team.'

'Your team?' queried Paddy. 'We don't work in teams here, other than for *Kegels*.'

'Eric's gone to Berlin,' said Dickie. 'He left early this morning. I know because I had to help him start his bloody car. Some damp had got into the carburettor – the devil's own job to start it!'

'Is he driving all the way to Berlin?' asked Cocky.

'No, he …' began Dickie.

'Have you signed the Official Secrets Act?'[9] asked Paddy. 'Seriously, chaps, we shouldn't be talking like this – where Eric has gone and how he's getting there. The next thing we'll be guessing is why he's had to go to RAF Gatow at short notice, and you never know who's listening!'

'No one's mentioned RAF Gatow except you,' cut in Dickie.

Paddy looked flustered.

'I've signed it, yes,' said Cocky. 'Don't tell me anything more. Except … when will he be back?'

The three men stood mute, unable to tell Cocky the official secret about when Eric might reappear to give him instructions for his first day's work – or the location of his office.

After bacon, eggs and fried bread, Cocky joined Geoff in a big room where a huge map was spread out on a table, fixed around the edges with drawing pins. It was old and full of pinholes, pencil marks and half-moon coffee stains. Geoff went over to a teleprinter that was pushing out a list of what looked like map co-ordinates, waiting while several pages appeared. The map on the table had a thick red line drawn on it to represent the Iron Curtain, dividing communist Russia and the Eastern Bloc countries of Western Europe. It was scattered with coloured pins. There was a cluster around Moscow, but no part of mainland Russia or the annexed countries was without a pin, apart from the mountainous areas – the Urals – and the Arctic wastes.

Cocky spent some time marvelling at the map. He loved maps.

Geoff's job was to use the co-ordinates to pinpoint places that had been mentioned by incoming signals over the past few days, and take a guess as to how they might be categorised. He explained that a yellow pin was a Russian research institute, a blue pin was a listening station, a black pin was an airport or airstrip, and a green pin was a location of not-yet-confirmed interest that would then be subjected to greater surveillance. If Geoff had absolutely no idea, he would put a white pin in the location, until someone could throw light on why it might have cropped up in a Russian signals transmission.

People sometimes wandered into the room on their breaks, to have a look at the map and make guesses about what might be happening where. It was very sociable, and Cocky started to enjoy it. He relaxed so much that he removed his jacket. Once the jacket was off, even more people came by, this time just to see his shirt.

'It's really intricate embroidery,' said somebody. 'Did you have it made?'

'No, I bought it in Changi, for a few dollars. I've got a pink one and a green one, too.'

'Ideal for the panto,' said another voice. 'Three colourful shirts would make short dresses – all the better to show the hairy legs beneath. "Three little maids from school are we" ... or something.'

'I'll be a schoolgirl,' shrieked someone else in a high falsetto, and everyone laughed.

'Me too,' said another, in a deep, gravelly voice.

'I've designed pantomime posters before,' Cocky admitted, and was immediately assigned that job for the Christmas pantomime. The last thing he wanted was to do was act on stage.

People were going back to work, still laughing. Cocky wondered if it was always like this, or only when Eric, the commanding officer, was away.

'These are all yellow pins,' said Geoff when the room had cleared. 'They are all research institutes, and all their co-ordinates have come from one, really long transmission that came through last night. See here, they all begin with N11 or OKB, followed by a number, and some of them have a bit of a description that the linguist has been able to add. We tape anything complex like this, and it gets listened over properly, to get the maximum information out.'

'N11-88', said Geoff, 'is one of the main ones. It's in Kaliningrad, north-east of Moscow. It's a place that used to be called Podlipki, but this transmission has used Kaliningrad.'

Cocky peered at the printout. He couldn't see the names Geoff had mentioned, as they were all written in a cryptic shorthand code that only some could decipher.

'One of the things the Russians do all the time is change the names of their towns,' said Geoff. 'Either to name them after someone who has done something noteworthy or just to boost the ego of one of the high-ups in the Communist Party.'

Dickie had wandered into the room and picked up on the conversation before drifting back out. 'It must be really difficult for the ordinary Russians who live there to suddenly find you don't live in Podlipki anymore, you've moved to Kaliningrad overnight, without even packing up your furniture – without even getting out of bed.'

Both men were studying the list too hard to laugh at Dickie's joke.

'OKB-456,' said Cocky. 'Looks like a made-up number.'

'The co-ordinates have that one sited at Khimki,' said Geoff. 'And the linguist has added a person's name in here, Gluschko, and has put "rocket engines" in here with a question mark.'

'So that one will be subjected to further surveillance?' asked Cocky.

'Yes', said Geoff, pushing a green pin into the location. 'It's only about 18km north-west of Moscow, and we think that if it's near Moscow it will be important. N11-10 keeps coming up, and not just in this transmission. We know there's a Russian scientist called Kuznetsov there, and the linguist has put "gyroscopic systems?" by his name, and he's also put "platform" and a query as he doesn't know if it's a station platform or another type.'

'It'll be a gyroscopic platform,' said Cocky. 'Eric showed me a new guidance system that uses gyroscopes. The platform is just the thing they're mounted onto.'

Geoff nodded his understanding.

Another question was forming in Cocky's mind. 'These descriptions seem to show that every small part of the rocket is being developed in a separate institute. Why don't they bring them all together into one place? Surely, that would be much more efficient.'

'It would be more efficient to have one big institute,' said Geoff. 'There's a manager of every N11 workshop, and every now and then they set up yet another one, or move one, or merge two together. One of our theories about why they don't have one big place is they are scared of being targeted. If they have one big institute, one factory, one mega-sized workshop, it would make an obvious target for the Americans to do to them what they did to Hiroshima.'

'So they don't put all their workshops in one place, in case the Americans smash them all up in one go? With a nuclear bomb?' asked Cocky in horror.

Geoff was used to the tension that exposure to the realities of the Cold War brought, and he was also used to witnessing the reaction of every new person as they realised it fully, too. He was still humming the tune from Gilbert and Sullivan's *The Mikado*, 'Three little maids from school'. He didn't answer Cocky's question.

'The Americans, and we British, must be so scared of the Russians achieving their goal, managing to build an atomic bomb,' said Cocky. 'The Russians must be living in fear, too.'

'It's not just the nuclear part the Russians fear,' said Geoff. 'The rocket delivers the bomb to its target. In Hiroshima, the bomb was dropped by men, by two pilots. They flew over the city, and then released the bomb when they were directly overhead.'

Cocky remembered the television footage of the enormous mushroom cloud rising above the Japanese city in 1945. He had watched it on the 7-inch screen of his mother's black-and-white TV, and despite the fuzzy reception and tiny screen, the destructive enormity of the bomb could not be minimised. It was a world-changing event, which had kick-started a new extreme of political tension, fear and frantic activity, just when the nations of the world needed to rebuild and heal.

'The V-2s have a direction-finding capability,' said Cocky lamely. 'So what the Russians are trying to do is, build a remote-controlled, nuclear bomb that can travel from one continent to another.'

Geoff grunted in agreement. 'The V-2's guidance system wasn't actually very good. They said it was, but if you look past the propaganda – and we've been able to get hold of experimental flight data here – it shows beyond doubt that the V-2's direction-finding was always failing. It relied on radio signals. They often lost radio contact at crucial moments.'

'That's hard to credit,' said Cocky. 'They really made out it was super accurate.'

'All bluster and bluff,' said Geoff. 'All part of the propaganda war, designed to mislead and frighten people. We've learned to see past the claims the Nazis made and now the Russians make about their successes. We take them with a pinch of salt.'

'So this one, N11-10,' said Cocky, pointing at the yellow pin on the map, 'is where the Russians are trying to rectify the problems of the Germans' V-2 rocket guidance system?'

'Just like the other workshops, it's only doing a small part of that. It's just working on the gyroscopic platform. They've given a guy called Kuznetsov the job. You watch, one day you'll see a Russian city "Kuznetsov-grad". It will probably revolve like a gyroscope.'

Geoff started pointing at the pins in the map, unpicking more of what he knew about each one. 'The whole guidance system will be being assembled somewhere else – we don't know where yet – and as each element comes together, the Russians will test it in an actual rocket. We think that there is a test site being used in a desert area – here.' He drew lines with his finger from the Russian launch site to the places that a rocket with sufficient range would be able to reach: 'London … Paris … Washington … New York …'

'The Americans must be petrified,' repeated Cocky. 'And so must our government, too. It would only take a few of these terrible bombs to wipe out the entire United Kingdom.'

'I'm sure they must be,' said Geoff. 'Petrified, I mean.'

Geoff suggested they stay calm and look at the facts. He wanted to prove his hypothesis that there was plenty of time for the political situation to change 'at the top'. He knew that the Russians had been trying to develop nuclear weapons for years before the Americans bombed Hiroshima. Back in 1946, President Truman put forward a non-proliferation treaty, called the Baruch Plan. In this plan, the US agreed to decommission all of its atomic weapons, on the condition that all other countries pledged not to even start production.

Cocky had heard of the Baruch Plan.

'All countries signed up to it except Russia,' explained Geoff. 'They thought it gave too much power to the USA. They've not signed up. They are accelerating their development of these weapons. That's really the crux of the Cold War.'

'These research institutes all over Russia, for instance. They've only just been set up, most of them. Some have recently changed from having another use – at least a couple of the places on this map were being used to develop chemical weapons.'

'That's reassuring,' said Cocky sarcastically. 'Only chemical weapons. I feel so much better now.'

'These institutes, or workshops, whatever they are, may not have produced anything at all yet,' said Geoff. 'We know they're struggling for components and equipment; we can tell by the number of requests flying around. Staffing too: most of them aren't properly staffed, and it's only recently that anyone has given thought to accommodation for workers.'

'This main one in Podlipki near Moscow, Korolev's one, it makes sense that that's where they will be co-ordinated from,' said Cocky. 'He's the chief.'

'Yes, we have come to the same conclusion,' said Geoff. 'Eric suggested that Korolev would direct operations from there.'

'How long would it take them, using all these set-ups, co-ordinated from Korolev's Moscow one, umm … N11-88, to produce a missile that could threaten America? Or the UK?' asked Cocky.

'Eric knows they have already tested one rocket; he tapped into signals from test launches in 1947.[10] But, we don't know any more about it as yet.'

'A test launch in 1947?' said Cocky. 'Did that follow on from the mass deportment of the German rocket scientists at the end of 1946? Eric told me all about it.'

'It's possible,' replied Geoff. 'The Russians couldn't have got that far that fast without help from the same Germans who had just developed the V-2.'

Marianna put her head around the doorframe and, with a smile, said to Cocky: 'There you are. Eric told me to tell you he has been called away urgently and won't be back for at least a week. He said to make yourself useful.'

'Oh, thanks, Marianna, I will do my best,' said Cocky.

Marianna disappeared, humming to herself, several people greeting her as she went past.

'There's one here, N11-4, with no co-ordinates given,' said Geoff. 'That usually just means that they haven't allocated it yet to a place; they're looking for a suitable site.'

'So we won't know where, what it's for and who is in charge of it, until another message mentions it?' asked Cocky.

'Exactly,' said Geoff. 'I put out words to listen for every day, so that one, N11-4, will go on the list.'

'What about this one, N11-88-5?' said Cocky. 'We have a Russian name, but no location. It just says "guidance systems" by the numbers.'

'No location, no pin yet,' said Geoff. 'Put it on the listening list.'

'There's one here that doesn't make sense,' said Cocky. 'It's N11-88 again,[11] but the location is way up here, 380km north-west from Moscow in the middle of a lake. Lake … Seliger. It must be a mistake.'

'Hmm, yeah, that looks like a mistake,' said Geoff. He studied the map. 'Lake Seliger is where the co-ordinates have it. But we know that N11-88 is in Podlipki, just outside Moscow.'

'Could it be an outpost of Korolev's place?' suggested Cocky. 'Like we have an RAF outpost at Gatow, on the East German border, Korolev might have an outpost on this lake.'

'Yeah, it's possible. It's a possibility,' said Geoff. 'They would call it a branch, not an outpost.'

'I'll put "branch?" on the listening list, then,' said Cocky, 'with a note to say it might be in connection with Korolev's place.' He wrote 'N11-88 – branch?' on the list for the radio operators.

'Eric has actually met this Russian, Korolev – he told me all about it,' said Geoff.

'Oh?' said Cocky, without giving away that he too had met Korolev, back in 1946. At that time, Korolev had been thin, with a haunted look. His scientific knowledge had seemed insecure, and rusty, and some basic developments in physics appeared to have passed him by. It was amazing to Cocky that this same person was now in charge of the complex operation they were trying to map.

Geoff was telling Cocky about Eric's experience of meeting Korolev. 'Eric was on a mission straight after the war to relocate a laboratory where V-2 rockets had been studied and developed during the war. Korolev turned up with some other Russians, interested in stripping the laboratory as well. Eric had a lot of chats with Korolev, and in Eric's opinion, Korolev is a bit of an enigma. He's a trained scientist, but his knowledge was way out of date. For example, he didn't then appear to know anything about long-range communication systems involving transducers. He even struggled with basic components, like thermionic valves.'[12]

Cocky had come to the same conclusions about Korolev. The man he had met at the Fraunhofer had been in recovery from a six-year imprisonment, and wasn't at the peak of his powers.

Cocky thought that instead of building up his own kudos at the Unit by mentioning that he had been with Eric on that mission, he would heroically keep an official secret.

Geoff was continuing with the reasons that suggested the Russians' nuclear capability was not imminent. 'Eric says, that trip to Austria, when he met Korolev close up, is how we know that a nuclear strike from the Russians cannot be imminent, but must be several years away.'

Cocky cast his mind back to the date of the mass movement, or abduction, of German scientists – October 1946, over four years ago. Surely, some of these workshops they were plotting onto the map must be productive by now. Eric himself had told him about the Soviet copies of the V-2 being mass-produced and supplied to the Red Army. He wondered if Eric kept worrying information from his staff, or if he too really believed that the Soviets' own nuclear project was in its infancy, and could therefore be neutralised before it had the chance to come to fruition.

Cocky wasn't as prepared as Geoff appeared to be to take Eric's word for it. He pressed a white pin into the middle of Lake Seliger, just in case.

'We've worked all through lunchtime,' said Geoff. 'Let's go and see if there's anything left.'

Most people had finished lunch and were discussing the *Kegels*, or skittles matches, planned for that evening. Teams and strategies were being considered seriously. Lists of names appeared next to the serving hatch. It was a good idea to put your name down for a team that had a chance of winning. Cocky scribbled his name next to Dickie Hunt's.

The skittles alley was in the backroom of a *Gasthaus* in the nearby town of Bückeburg. It was a long, low room accessed from a green wooden side door. There was a story that a tunnel linked the alley to one of the rooms at the Signals Unit, forming an escape route in the event of an attack.

Two other men joined Dickie and Cocky to make up a team, and the others explained the game. 'There are nine pins set on the diagonal, each player gets fifteen goes,' they said, pints of German lager in hand, while someone set up a new scoreboard.

'It's an incredibly long lane,' said Cocky. 'And the balls don't look very round to me.'

'They're not spherical balls,' said someone. 'They are practically prehistoric. And the lane isn't the most impressive playing surface. It's really flat to start with but towards the back, it's badly warped and if you hit the worst part of it, you won't hit the target in a month of Sundays.'

'There's a badly worn area about halfway down,' said Dickie. 'You can see it if you squint.'

'And the edges – avoid those if you can,' said Paddy.

'Avoid going straight down the middle, too,' said Dickie.

'And don't throw the ball up so that it bounces,' said Geoff.

'The best tactic is to try to get your ball weaving back and forth so that you come sideways on to the skittles,' added Dickie.

The game sounded practically impossible. Cocky took a few slurps of beer to steady his nerves.

Air Scientific Research Unit, Obernkirchen, 1951

Dickie had a twinkle of amusement in his eye. 'It's great fun,' he said with conviction. 'Let's get started.'

Cocky's first ball was smooth and worn, and egg-shaped. It skittered lumpily down the lane, veered to the left, and stopped about 10 feet short of the pins. People with more familiarity with the idiosyncrasies of the lane managed impressive shots and strikes that seemed to defy logic and make nonsense of the rules of physics.

'Newton's Third Law!' shouted Paddy every time he made a strike, which was often. 'Equal and opposite: action and reaction!' he screamed as the pins went flying yet again.

'I'd like to programme this ball to avoid the warped bit, then start weaving back and forth about a foot from the target,' said Dickie.

'It could be done,' grinned Cocky. 'A tiny analogue computer,[13] with the desired trajectory programmed in, and fixed inside the ball. On a gyroscopic platform.'

'The other teams would be absolutely bloody amazed,' said Paddy.

'They'd be amazed for five whole minutes,' said Cocky. 'Until they had a look at our ball and worked out how we'd done it.'

'Then we'd get banned,' said Geoff. 'Which would be a shame because this game is the best fun you can have without women around.'

The noise level had been steadily rising and the floor was sticky where beer had been spilled. The low ceiling made the air warm and humid. Another team was way ahead, and apparently this was normal.

'Atko's team are unstoppable,' declared Paddy. 'They've won every match for weeks. No one else is even close.'

'Sorry, I've not been very good at it,' said Cocky.

'You'll get there,' smiled Dickie. 'To be honest, no one really cares less. It's just a night out, away from the Unit, the best we can do for ourselves away from home.'

The word 'home' had a sobering effect on the previously ebullient players. As the scores were read out to jeers and shouts of 'Fix!', Cocky slipped out into the cool night air. Dickie and Paddy weren't far behind, so the three of them walked back together.

'See you in the morning,' said Geoff.

'Oh, I'm looking at parabolic antennas with Paddy tomorrow, first thing,' said Cocky. 'I'll look in on the map room at some point during the day; see if there's anything I can help with.'

'Fine. G'night,' said Geoff.

The next morning, Paddy was nowhere to be seen, probably sleeping off the night before. Cocky wandered into the main room, where six people, all wearing headphones, were typing – their expressions as intensely focused as they had been at the *Kegels* game. He watched and listened for a long time before anyone had the chance to stop and talk to him. Eventually, a young man with a round white face and glasses stopped operating his machine and took off his headphones. He sat still for a while, recovering as if from running a race. Then he glanced down at the notes he had made in response to the signals transmission.

Cocky went over, to see what he had written. There were breaks in the narrative because the operator had needed to keep retuning his radio receiver to an ever-changing frequency. He had done his best to find it again as soon as it disappeared, but this could take seconds or even minutes.

'They were changing the frequency every few seconds,' the operator explained to Cocky. 'When they do that, I sort of know that this is an important one.'

'May I?' asked Cocky, and the operator showed him his transcript of the signals he had received.

'What does "Lucky"[14] mean here?' asked Cocky. 'You've got the word "lucky" about four times, and it doesn't fit with the rest of it.'

'It'll be a code name,' said the young man.

'And there are a few of the N11- numbers I was looking at yesterday with Geoff,' said Cocky.

'Yes, those are the Russians' "research" places,' said the operator. 'Like everything that comes out of Russia, "research" is a bit of a euphemism, and it'll be where things are being designed, developed, tested – it's never just a bit of harmless research.'

'Can you tell me as much as you can about this transmission?' Cocky asked the young man, pulling up a chair. 'It's ok; I've signed the Official Secrets Act.'

Air Scientific Research Unit, Obernkirchen, 1951

The young man moved his glasses, which had slipped down, back onto his nose. 'The timing of it is important. Getting this first thing in the morning here means that it was being sent as soon as operations begin, in Moscow.

'If you look at it, I've translated it from a form of code that is used only by the top officials. There are expressions that translate only from the most correct form of Russian – no shorthand or slang.'

Cocky was impressed. His rudimentary knowledge of Russian could not have made this distinction.

'It's mentioning a research and development project,' said the operator, 'with the code name "Lucky".'

'What does "black bread" mean, in this bit here?' said Cocky.

'They often use foodstuffs[15] to indicate different groups,' said the operator. 'So "black bread" would most likely mean the East Germans, because they seem to like rye bread.'

'And what about this strange word?' asked Cocky. '"*Falsch-mann*", it looks like. Then you've put "FM" for short a few times. Is it "frequency modulation"?'

'No, that's not what it means. FM is their word for the President of the United States,' said the operator. 'His name is Truman, so the Russians have played on "True-Man" and have corrupted it to a German word, *Falschmann*, i.e. "False-Man". We've had that one through a lot, and we agreed it is code for Truman.'

'Apart from underlining that the Russians don't trust the Americans, what do you think we are looking at here?' asked Cocky.

'It's not the only message we've had recently on this subject,' said the young operator. 'Our accumulated intelligence seems to indicate that the Russians are planning a new generation of rocket, which will be able to carry a nuclear warhead, and they are going to mass-produce them and site them somewhere in the eastern part of Russia or in East Germany. As near to Western Europe as they can get. The rest of the message is estimating costs and development time.'

'Could "lucky" mean "Lucky Strike?"' said someone else. 'As in, the American brand of Lucky Strike cigarettes? Is it a strike against America?'

'Or it could have something to do with the number seven,' said another. 'Lucky Seven.'

'I don't know if "seven" is lucky in Russian,' said someone else. 'I think it might only be a lucky number for us Brits.'

'It is,' shouted another voice – accented and unmistakably Russian[16] – from across the office. 'Seven is lucky in Russia too.'

'Seven might just mean the seventh in the series of rockets they've been working on,' said another. 'We've had R-2 and R-5; this could just be R-7.'

'What happened to R-6?' quipped somebody.

The idea that the Russians hadn't got far with setting up their workshops now rang hollow with Cocky. A seven in the code name, where earlier iterations had lower numbers, looked like a series, and if the series had reached number seven, then the Russians were much further on with versions of their rocket than Eric had admitted to his staff.

'R-7,' said Cocky. 'Sounds serious. And, does it say who has sent this message?' he asked the operator.

'It's the current code sign for Korolev's office; they call them "bureaus",' said the operator. 'And anything from his office has to go to GCHQ, no matter how innocuous it might appear.'

'Well, this one's not even pretending to be innocuous,' said Cocky.

'We send all our intelligence through to GCHQ every day,' said the operator. 'It's evaluated by us, of course, so we give as much background as we can, and we'll put in our ideas about interpretation on the daily report. So I've included this discussion we've just had about this word "lucky" in the report.'

'Flown out every day from Bückeburg Airfield near here, to GCHQ,' said someone else. 'We have to put the things we want them to notice first on the top of the pile.'

'And does anyone tell Eric?' asked Cocky. There was a silence. The matter of keeping Eric in the loop was obviously a sore point with the operators.

'There is so much information', said someone eventually, 'that no, we can't always update Eric. If he's not here, it gets swamped by other stuff, and by the time he's back, this R-7 "top secret" will be old news.'

'Seriously,' said someone else, taking off his headphones and swivelling his chair around to face Cocky, 'we are finding out all the time about Russian activities that may or may not be a threat.' Bazza over there, for example, was listening in late one night – weren't you Bazza? – and he picked up a long conversation in ordinary Russian speech – very unusual for it to be not coded at all – which seemed to be about a "city" being built in a place near Yekaterinburg.'

Bazza was eager to take up the tale. 'The thing about this city was, huge numbers of German prisoners of war, and ordinary Russians, were apparently being used to build an enormous plant – no mention of what they were making there. A workers' city from which no one can ever leave.'

'GCHQ wanted more information,' said someone else.

'But we've never been able to find any more information,' said Bazza. 'The number "94" came up a lot, and listeners are on permanent alert for that number coming up, but it never does.'

'Atomic number 94?' tried Cocky. 'Plutonium?'

'Probably,' said Bazza.

Cocky wanted to return to the transcription of the transmission about the R7s. In particular, he wanted to know what the Russians' estimate of the time their new rocket would take to develop to the test stage. This estimate, he reasoned, would give the American and UK governments an idea of how much time they had to diffuse the extreme political tension.

Paddy's Antenna-Turning Mechanism

Paddy Engelbach wandered into the listening room and joined in for a while with the general discussion about the unknown city, before asking Cocky to come and look at his design for the turning mechanism for a parabolic antenna. The mechanism had to enable the antenna to rotate automatically in the direction of radio signals[17] coming from anywhere of interest, so that the antenna could amplify and receive information. Paddy had come up with a design that he wanted Cocky's opinion on.

Cocky grabbed his box of British Thornton drawing instruments and followed Paddy through a maze of corridors to a quieter suite of rooms, well lit and equipped with drawing boards.

'I relocated the drawing office myself,' said Paddy proudly. 'I can't get much done when everyone's pulling me into discussions, so I set myself up here. It's got good light and it's quiet.'

Cocky thought that the very looseness and informality of the Unit was what made it a creative and interesting place to work. The discussions that flew around the offices were light-hearted, and anyone felt they could contribute without being censored. It was a great way of pooling expertise, helping each other reach better decisions, and make sense of barrages of information.

'What's the antenna for?' asked Cocky, and Paddy began his explanation of the brief he had received from GCHQ, which wanted a set of bigger and more powerful dish-shaped receptors that could rotate and scan for the best and clearest signals.

A few weeks later, Paddy had perfected the antenna design apart from some gears, which the rotating base needed to move through its arc. Cocky remembered that a designer at the Bristol Aeroplane Company in the UK, where he used to work, had invented a similar gear drive for another application. He offered to go and get one made in Bristol, to Paddy's exact specification. Eric was persuaded that this was the right course of action, even authorising Cocky to take a few days' extra leave in Bristol, because he remembered that was Cocky's home town.

Cocky's 'office' and 'team' at the Unit never actually materialised, but he found that this gave him the freedom to simply follow up any piece of information that seemed to be important. Absorbing work and *Kegel* nights had made the last few weeks slip by, but Cocky had received a couple of troubling letters from his sisters. He had good reason to visit his mother in Bristol.

Eric drove Cocky to Bückeburg Airfield to meet the plane he had organised, only for him to find that it was the distinctive Anson aeroplane belonging to the commander-in-chief. Cocky was the only passenger and he was invited to sit in the navigator's seat in the cockpit. Eric was like a little boy who had bought his friend a present he couldn't wait for him to open, as he showed Cocky around the Anson and settled him next to the pilot.

'I don't need a navigator for such a short trip,' said the pilot. 'But just in case anyone asks, that's what you need to say you are.'

'Thanks, yes,' grinned Cocky, as he buckled up the various harnesses that kept him in the seat.

'More comfortable seat than that old Bedford,' quipped Eric, referring to the expedition the two men had made together over five years ago. 'Don't fall asleep, now.'

'I won't,' said Cocky, eyeing all the buttons and levers that controlled the plane.

Eric jumped down from the cockpit and strode over to his car, which he had parked near the runway. Eric never stayed still for long and Cocky had hardly seen him since the pleasant evening of his arrival. Cocky had been planning to re-read his sisters' letters during the flight, but this chance to get near to the experience of piloting a plane was too good to miss. He watched Eric's car disappear as they taxied for take-off. Just as the heavy Anson rose into the air, he spotted a black SOXMIS car parked a couple of fields away. One of the occupants had binoculars trained on the cockpit.

'SOXMIS,' said Cocky to the pilot. 'Soviet Military Mission. Let's see if I can spot its number plate.'

The pilot grumbled that SOXMIS vehicles were always snooping around. Part of the post-war agreement between the Allies, including the Soviets, was to allow visits to other Allied units. Designed to prevent espionage by removing the need for it, and build trust between the Allied forces, the Soviets were the most prolific users of this provision for free movement.

'I can't see its number,'[18] said Cocky. 'It's facing the wrong way.'

'I don't know whether we bother to send cars over to their side,' said the pilot. 'I don't think for a moment that we trust them more because they can drive up and down our roads, taking bloody photographs.'

'They're taking photos now,' said Cocky, as the ground below them grew too distant to make out any further details, and the wide open sky became their view.

The flight was breathtaking but short, and soon the pilot was guiding the Anson down onto the airfield. Cocky unbuckled his seat, thanked the pilot, and then headed for the familiar bus that went into his home town.

Home Rescue Mission

In a flat in Bristol, Cocky's mother was suffering the effects of another of his father's failed business ventures. Iggy had taken his frustrations out on his wife, beating her badly. It was always worse when he had been drowning his sorrows and spending the last of the food money in the pub.

Marjorie's letter to Cocky outlined some of the treatment that she had tried to prevent by moving back in with her parents. Marjorie had had to move out again because their father was starting to turn on her. All the siblings worried that one day they would find their mother badly injured, or worse. They were running out of ideas about how to protect her. The local police were bemused when Marjorie had asked them to help protect her mother; as far as they were concerned, this sort of physical abuse was normal, a private matter between a husband and wife. The wife was, in English law,[19] the husband's property, to do with as he wanted.

Before going home to his parents' flat, Cocky called at a letting agent's office. The pleasant lady there helped him find a small, neat flat on the other side of Bristol. He paid the deposit and put the tenancy in his mother's name, arranging to pick up the key in two days' time. He found a phone box and telephoned his brother and sister. They worked out a plan to take their mother out for the day, while Marjorie packed up and moved her belongings. They would return their mother at the end of the day, not to the home that held danger and so many bad memories, but to the new flat, safe and clean, and full of sunlight. It was of paramount importance not to explain the plan to their mother in case their father wrung the details from her.

Cocky had saved enough to pay the deposit and the rent on the flat for a year. This was the best use of his savings that he could think of. All that remained was to give his mother the tactical day trip. His thoughts turned to cars, and he felt the best idea would be to borrow his brother's Alvis and take his mother to Caerphilly in Wales. They could look round the medieval castle, buy some cheese, and return in the evening to the new flat, which Marjorie would have made homely.

All the time that Cocky was negotiating with the Bristol Aeroplane Company about the gearing mechanism for Paddy's antenna, he was thinking about the escape plan. The main problem was being sure that his father would stay out of the way for the day. Cocky knew his father had a mistress who lived at Weston-super-Mare, and the idea that this could be useful took hold in his mind.

'Hello, Mother,' he said, arriving on the doorstep of his parents' flat.

'Ooh, my Neville, is it really you?' said his mother with delight. 'Would you like a cup of tea?'

'A cup of tea is just the ticket,' said Cocky. 'Where's Iggy?'

'At the pub, or maybe the club. I'm not sure,' she said. 'He won't be back till late – of that we can be certain.'

'Let me look at you,' said Cocky, holding his mother and taking in the state of her face and arm. To him, her face seemed strained, and she looked a lot older. 'Are you still in pain? John told me what he did.'

'Oh, he didn't mean to hurt me. He was looking for his tobacco, which I had stupidly moved away from its usual spot. He was asking me where I'd put it.'

'Asking you with his fists?'

'Well, I really shouldn't have moved it. He always wants a smoke when he gets back from the pub. I should know that.'[20]

'How about a day out, just you and me?' asked Cocky. 'On Wednesday, we could drive out to Wales.'

'Wednesday, I need to make Iggy's dinner and tea,' she replied. 'There will be hell to pay if I don't do that for him. You know that, son.'

'But, would you like to do to that, Mother? If you didn't have to worry about Iggy, would you like a drive out with me? To Wales?'

'I don't know ...' she began. Then, 'Alright, yes. It would be lovely. To go somewhere in a car with my handsome son. Have you got a car?'

'I'm borrowing John's,' said Cocky.

'It sounds like you've already planned this,' said his mother.

'I have planned it. I've not got a lot of leave.'

'You've already asked John about borrowing the car ... but I'm not sure if I can have a day off from my duties yet.'

'Please, Mother. Please say yes. For me. I've only got a few days, and I don't want to spend it here in this flat. I'd love to have a drive down to Wales, with you in the passenger seat.'

'But it's afterwards. I'm sure that you and I would have a lovely day, but it's what would happen afterwards, when Iggy doesn't get his dinner and tea, he'll go mad. He'll probably go straight down the pub, drink himself blue, then ... you know ... he'll make his displeasure known.'

'Mother, trust me. It will be fine. It's going to be perfectly alright for you to come with me on Wednesday.'

'I don't know. I think you'll have to invite Iggy as well,' she said.

Cocky considered this. It would be a huge risk to invite Iggy. If he said yes, and came with them, it would spoil the whole plan to move his mother permanently out of harm's way. If, however, Iggy saw the opportunity to visit his mistress in Weston-super-Mare for the day, at least then his mother would not be dreading another beating. Iggy might even be in a good mood. It was a fine line to tread between deceit and openness – a fine line that could have terrible consequences, particularly for his mother, if it went wrong.

'I'll ask him,' he said finally.

'And we'll hope that he has other plans,' whispered his mother.

'More tea?' asked Cocky, as he heard a key in the door.

'Whose is this?' shouted Iggy's voice, kicking Cocky's bag, which he'd left in the hallway.

'Mine,' said Cocky. 'Hello, Father.'

'Hello, son. Have they thrown you out of the RAF? Weren't you any good?'

'I've got a few days' leave,' said Cocky.

'Bloody soft, the lot of them. Leave, eh? Never had leave in my day. Never had a day out of the trenches on the front line.[21] You lot don't know you're born. There isn't even a war on!'

'There is a sort of war on – a hidden war,' said Cocky.

'No such thing as a bloody hidden war!' ranted his father. 'What a load of rubbish. You're a pointless waste of space, just like your mother. Why don't you get back to whatever you call work?'

'Mother and I were hoping to have a day out in Wales on Wednesday ...' Cocky's heart was somewhere near his boots, yet pounding in his ears. 'And we'd like to invite you.'

'What's in Wales?' shouted Iggy. 'Other than pixies and ghillies and bad weather?'

'There's a lovely castle,' said Cocky's mother bravely.

'A castle. Well, well, a castle is there? And what good is a castle, may I ask? Will a castle pay the rent for this place? No, it won't. Is a castle going to put food on the table, clothes on our backs? I have never known a castle to do all that. I can't think of a more pointless waste of time than going all the way to bloody Wales to see a bloody castle!'

'I've heard there are jobs in Weston-super-Mare,' said Cocky, risking everything just to mention the place.

'Weston, eh?' said his father. 'Yes, well, Weston is a seaside place; I could pick up something seasonal.'

'Painting boats?' suggested his mother. 'You used to be a merchant sailor.'

'Shut your face, woman,' shouted Iggy. 'I will paint boats if I decide to paint boats and not on your damn fool suggestion. If you make any more comments along those lines I will not be held responsible for my actions!'

Cocky and his mother did not dare to exchange glances.

'You two go to damn fool Wales and ponce about seeing stupid castles, and I will earn money and show the lot of you just how many people want to take me on,' said Iggy, reaching for his whisky bottle and pulling it close. 'And I might find the time to call in on a friend I've got there too.'

Cocky breathed out. He hadn't realised he was getting dizzy from holding his breath while his father opted out of the crucial outing to Wales. He sat back down at the kitchen table, trembling.

'What's for tea, woman?' demanded Iggy.

On the plane back to Bückeburg, Cocky had the new gearing system in his bag for Paddy's antenna, and he had two ticket stubs from Caerphilly Castle, which he and his mother had fully enjoyed on their day out.

The day had been magical: the Alvis had run like a dream, the weather was perfect and his mother seemed to shed at least ten years of her age as she sat in the passenger seat, enjoying her treat and the precious time with her son. She had talked about holidays she had known as a child, and Cocky realised that all her good memories were from her own childhood – there were none from her married life. On the way back, a smelly package of crumbly Caerphilly cheese on the back seat, Cocky had told her that she would not be returning to the flat where she had set off from, but would soon be installed in a new, lovely flat of her own, where her other children would visit her often, and where she would be safe from Iggy.

She sat, stunned, in the passenger seat, tears coursing down her face. 'What will Iggy do?' was her first question. 'He needs me. He can't cook.'

'He will be fine,' said Cocky, who hadn't considered his father at all.

'He will find me,' said his mother. 'He's got eyes and ears all over Bristol. He will find me and he will kill me, for leaving him.'

'He won't. The others will keep your secret, and keep you safe.'

His mother had eventually accepted her children's solution for her safety and happiness. She loved the new flat, walking around it touching the clean, shiny surfaces. She was delighted that Marjorie had moved her things into it, and had been working all day to create a home.

During the flight, Cocky closed his eyes. He enjoyed replaying his mother's reactions to her new home, her new life. It felt like the most satisfying thing he had ever done.

The plane touched down, and Cocky gripped his bag, fishing out his passport and his jacket. A single car was waiting on the tarmac. A long, black car, with curtained windows and a big yellow plate with a black number.

Abduction from Bückeburg Airfield

In the seconds it took them to shove him into the back seat of the SOXMIS, Cocky's heightened awareness for car makes and models kicked in. He noticed it was an adapted Chaika 13 – a Russian copy of the Packard that Eric drove around in. There was a diplomatic number plate – a big black 43 on a yellow background – next to a red Soviet Military Mission badge. The two black-clad heavies sat themselves either side of him. One of them threw Cocky's bag in the footwell while the other tied his hands behind his back with a length of scratchy rope. Cocky remembered to push outwards with his wrists so that the rope would not be too tight, but Russian brute force easily overcame this ploy. There wasn't a blindfold, possibly to avoid attracting the attention of the still-present American military guard.

Cocky sat stock-still with shock, his heart pounding, his mind racing. He didn't know who was due to pick him up from Bückeburg Airfield. It might be Eric or Marianna, or any other member of the Unit. The only person guaranteed to notice that Cocky hadn't arrived was Paddy, because he was waiting for the antenna gearing mechanism that Cocky had in his bag from Bristol. The mechanism was about 4 feet long – 2 feet long when folded – and was a prototype, but Paddy couldn't complete his design without it. He would surely raise the alarm when Cocky didn't turn up.

The car accelerated past the guards at the airfield without raising their interest, and Cocky cursed the slackness of American post-war sentries. Even in his panicked state, he also found time to curse the curious arrangement whereby Russian SOXMIS (and British BRIXMIS) vehicles could roam freely between the territories held by the Allied powers in Germany.

SOXMIS was an arrangement designed to generate trust between the Allies.

SOXMIS was the perfect cover for a kidnapping.

The driver had not turned around, clearly intent on getting smoothly away from all military paraphernalia and people who might just realise that the British RAF officer who had just been deposited on the runway had disappeared. There would be no car chase like there would be in a film, thought Cocky. He kept still, eyes focused on the neck of the driver in front, noticing, in his heightened sense of awareness, his serious acne scarring.

Something moved in the deep footwell. It was a dog – not as large as a German Shepherd, but a sniffer dog just the same. It was sniffing Cocky's bag with interest. Cocky thought about his bag. The antenna-turning mechanism was lodged at the bottom, beneath his clean clothing, lovingly laundered by his mother. In a side pocket, nearest to the dog's nose, there was also a large parcel of Caerphilly cheese his mother had pressed on him, packed up in greaseproof paper.

The curtained windows blocked all but the front windscreen – a view filled mostly by the driver's shoulders, neck and head ... which looked strangely familiar. The hair escaping from the cap was straggly and brown, nondescript except for the growing feeling that he had seen it before. The distinctive acne scarring was something he was trying to dredge from his memory. He mentally listed the facts of his situation, which helped him not to panic:

- In a SOXMIS vehicle, number 43, with two heavies on either side.
- A driver he was sure he had seen before.
- His bag, full of clean clothes, smelly cheese, and the gearing for an antenna mechanism.

- The car moving along a straight, fast road, direction unknown.
- Hands tied, but no blindfold. No assault. Not been stowed in the boot.
- A sniffer dog. Asleep.

Cocky's thoughts turned to the Unit, and the likelihood of someone raising the alarm. Eric, perhaps, or Paddy. The informality of the Unit belied the fact that it took seriously its rigorous security, and his name would be on a list of people expected to arrive that day. At some point, the gatekeeper would report his non-arrival. *Then, Eric will do something. I don't know what. He will do something.*

The car turned right. The driver's face moved slightly and a hooked nose profile was briefly visible.

Mitya, thought Cocky. *Mitya is the driver. His nose, his shoulders, his acne scars. He must have identified me, and he is now facilitating my abduction.*

The dog woke up and growled, sensing Cocky's slight start on recognising Mitya. The Russians began exchanging some desultory conversation. Cocky listened but was only able to make out one word in ten, especially as they used code words, slang and nicknames. The SOXMIS slowed and stopped at a checkpoint, where no one bothered to look inside the vehicle. Cocky wasn't sure if they were in the Russian zone or East German territory, because either of those would explain the languid wave-through for the driver.

Now the heavy car's engine was protesting as it laboured up the metalled mountain road, bends getting ever tighter as they progressed. Cocky resisted the urge to advise Mitya on how to manage the bends better. He just sat in the back, wedged between the large, sweaty men, determined to take in as much as he could of what was happening. They were clearly climbing a very steep mountain. No one spoke. The dog was asleep. Cocky's ears popped with altitude. He wanted to pull his earlobes to help them adjust, but his wrists were still tied. The sound of a steam train crept into his consciousness, barely audible at first, then louder and louder, until it came into view above them on the mountain – a narrow-gauge steam train with black-and-red livery.

A conversation he had heard at the skittles game in the *Kegelbahn* came to him in fragments, which he now pieced together. The previous year, Eric had arranged a skiing trip for six people from the Unit to the Harz Mountains. He had only wanted good skiers, because the terrain wasn't suitable for beginners, and the man who had been chatting with Cocky at the *Kegels* game had been proud to be one of the party of six people chosen. Eric had driven them himself, in his Packard, to the Harz Mountains. When all six of them were at the top of a ski run, he had warned them not to go off piste down the other side of the mountain because they might end up in the GDR, which was Soviet East Germany. The man from the ski party had mentioned a steam train on the mountain. He had suggested they take the train to the top, but Eric told him it was for military supplies only, for the Russian intelligence agents to do their listening-in to West Germany. Eric had pointed out the huge antennas erected at the top of the high-security mountain for exactly this purpose.

If this was the mountain that Eric had driven the ski party up the previous year, they were probably on the exact same road he used, and his Packard would have had the same difficulties with the sharp bends as this car was having. How had Eric managed that, with Russian security everywhere? Perhaps he had attached a BRIXMIS plate to the vehicle. Or, more than likely and more typically of him, he had probably trusted his own audacious confidence. The ski trip had been a great success and had gone down in ASRU legend as one of Eric's best ideas. It even inspired a guest spot in the Christmas panto, with a group of lost skiers appearing in a bemused state at Cinderella's ball.

Cocky focused hard on remembering the name of the mountain. He couldn't easily summon it up from the jumble of his memory. He went through his mental cache. *I am in the Harz Mountains, on the military one that has a train, and Mitya is the driver. This mountain has a billet for Soviet and possibly also East German spies, and it will definitely have a lot of radio equipment*, he recalled more clearly now.

'Brocken,' said one of the heavies to the other, as part of a sentence that Cocky didn't understand.

The Brocken – the name of the mountain – pinged neatly into his brain.

The Brocken

By the time Cocky was unloaded from the car and shown into a shared bunkroom with four Russians, he had a formed a hypothesis about exactly where he was. Because of Mitya, he also had an idea about why he had been targeted. His wrists were untied and his bag pushed at him. All four of his bunkroom companions were armed. He couldn't leave the billet hut without being shot. The dog, too, was stationed by the hut door, and it didn't take its ears off Cocky. It was listening to his every movement, occasionally backing this up with a glance.

Plates of food appeared at the door, with one for Cocky, and he was surprised to see that the person delivering them was Mitya himself. It was snowing on top of the mountain and a gale was blowing, so Mitya couldn't stand there for more than a moment.

'It's not chicken stew,' he said clearly, with a small, self-conscious smile. 'It's cabbage soup.'

'Why am I here?' Cocky asked him. But Mitya couldn't risk a conversation. He delivered the food with a basket of bread and shut the door firmly.

The other Russians ate their meal, drank vodka and started to play cards. Cocky was still hungry so he reached for the cheese from his bag. The guards reacted immediately, training their guns on him. Cocky brought the cheese into their view and broke off a lump to eat on his bunk. The guards lowered their weapons uncertainly. Cocky broke off some more of the cheese and shared it around.

Caerphilly cheese. Only yesterday, it had been a thing of joy, a gift bought by him for his mother to celebrate her first day of freedom in the whole of her married life. And now, the cheese had another role. It was helping him to deal with captivity, with hostility, with fear. The guards expressed their appreciation and indicated that Cocky could play cards with them. So, he played cards and ate cheese, if only to while away the time, and to build bridges where he could.

In the four or five days he was kept there, he never once was left alone, or had the opportunity to find a radio set he could use to contact the Unit, or Eric. From the window, he could just about see a building (see plate 15), like one of the tenement blocks he had lived in while growing up in Bristol. On top of the building was an enormous antenna, stretching far up into the sky, its apex obscured by the window frame. This massive listening post confirmed to him that he was on the very summit, just a few kilometres into East Germany, and therefore Soviet territory. It felt good to be sure of exactly where he was, but he had no way of using the information.

'You are Cock-sky?' asked one of his card-playing companions on the third day. 'They are looking for you, your British friends. They have no idea why you didn't arrive at Bückeburg Airfield. They think you must have stayed in the UK.'

'They think you have stayed with your mother,' said another Russian. 'In the UK.'

'They will stop looking for you, maybe tomorrow,' said the fourth Russian. 'Then we can go.'

Cocky swallowed hard. He was relying on his colleagues at the Unit not to make lazy assumptions about why he hadn't materialised at the airfield. He wanted Eric, in particular, to realise that he had been abducted, but was Eric even aware of his disappearance? He may not even have been told. No one else at the Unit had any idea that Cocky could conceivably be targeted by the Russians. He hadn't really had time to become a fixture at the Unit, and without an office or a set group of colleagues, people just wouldn't notice his absence or detect any sort of foul play. He wished he had opted to play a part in the pantomime. At least this would have ensured he wasn't forgotten.

'Go where? Where are we going?' he asked.

But the Russians just cut the cards and dealt them round again. Cocky could hardly play cards. He knew that Eric was the only one who would definitely not believe he would opt to stay in the UK. Flight records would surely show that he was aboard the military plane to Bückeburg. But if he was still marked down on the flight records as the navigator, not a passenger, he might not be missed at all. He regretted now going along with the small white lie about being the navigator.

And Paddy. Paddy would know that Cocky would not leave the antenna-turning project without a word. He could deduce that something was not right, and would surely be the one to tell Eric. It was a long shot. By now, Paddy could have found a way around the antenna-turning mechanism problem.

Cocky tried his best to fight down the waves of anxiety that kept coming over him.

The next day, the Russians took Cocky on the small steam train back down the mountain and then drove him to Berlin Airport, where they put him on a civilian flight to Leningrad. At Leningrad's airport, a driver was waiting in another diplomatic car. There began a six-hour drive on a straight road through the Russian countryside, stopping only at a town called Novgorod for fuel and food.

Cocky knew by now that he was the subject of a plan and that he would have no chances to escape. Thankfully, inside the locked car, his hands were free, but there was nothing to do except look out of the window at the unending landscape, its few isolated houses, the occasional countryside inhabitant, each village the same as the last. The landscape was flat but had a certain unspoiled beauty.

The Russian driver was silent, not sharing even one word. Toilet stops were conducted with efficiency, and Cocky was handed bread at regular intervals. His brain was spinning with anxiety and indignation, willing his colleagues at the Unit not to give up on him, not to assume he had stayed in the UK. He didn't want to become one of the people who 'disappeared'. He wanted Eric and Paddy to receive the telepathic messages he was trying to project. It was a bit like praying:

Abducted from Bückeburg Airfield, taken to the Brocken, now in Russia. Don't give up on me.

Chapter 4

Gorodomlya Island

Helmut and Irmgard

As twilight fell around the car, Cocky became aware that he had been asleep, and that the car had now stopped at a small jetty surrounded by armed guards. The jetty stuck out into a lake and there was a small steamship moored alongside it. Russian guards were checking identity papers, including his own passport. He noted with dismay that his military passport was not returned.

The steamer had been held up in anticipation of his arrival, and the skipper was anxious to be off. Cocky stepped aboard with his holdall and was suddenly alone on the boat, in the dark, with just the sound of the engines gently propelling him across inky black, smooth water. The boat arrived at an even smaller jetty on the other side of the lake, with a guarded gate and a high, barbed wire fence stretching into the distance.

Once inside the fence, all vigilance around his possible escape was relaxed, and Cocky realised that the fence must go all the way around a defined area, within which he was a prisoner. He carried his own bag, and the skipper escorted him through some woodland to a stone-built house, set apart from a long row of smaller houses. The limewash on the outside walls made it glow slightly through the gloom. After tapping on the door, the skipper disappeared back into the night.

A German housewife opened the door, and a warm light flooded out into the darkness. The woman's serene face broke into a wide smile, her halo of dark blonde hair backlit, and for a moment, Cocky believed in angels. It was so unexpected to have a benevolent face beckoning him into an actual house – not a barracks or a bunkroom, but a real, family home.

'Irmgard?' came a man's voice from behind the woman, and Cocky noticed her dampen her smile immediately.

'Who's at the door at this hour?' asked the man in German. Two children, a boy and a girl, came running out too, and were friendly and curious.

'Our visitor, your new employee. We've been expecting him, remember, Helmut,' said Irmgard. 'Come in, come in,' she urged.

Cocky stepped inside. Irmgard closed the front door, and showed him into the family's small sitting room, where he noticed a large wooden radio[1] set on the sideboard.

'Are you hungry?' she asked.

'Yes,' replied Cocky, suddenly realising how very hungry he was after a diet of nothing but cabbage soup and bread for the past week.

Irmgard and the children went into the kitchen to prepare some food,[2] while Helmut settled into his armchair opposite, and appraised Cocky. 'You understand radio wave guidance systems?' he asked bluntly. 'For rockets?'

'Ummm, not really,' mumbled Cocky.

'I've heard differently,' said Helmut. 'Come now, what is your experience with V-2 rockets? You may know it as the A4, and the Russian system has them as R-2 … well, they're up to R-7 now, but they are all the same technology. Come on, you know about these, no?'

'No, I'm an electrical engineer – I've worked on aeroplanes.'

Cocky was tired, hungry and disoriented after his journey. Helmut's barrage of questions felt like an attack. He really needed the food and rest that Irmgard was trying to allow him.

'I was involved with developing the V-2 at Peenemünde,' said Helmut pompously.[3] 'Very involved. Particularly the exhaust nozzle engine booster. That was mine.'

Cocky knew that Peenemünde was on an island in the Baltic Sea between Germany and Poland, where the Germans had developed and tested the V-2 silent flying rocket bombs used against the UK in the latter stages of the war. He knew the British had successfully attacked the rocket-building facilities at Peenemünde from the air in Operation Crossbow in 1943. It was hard for him to talk about events from the war years with a senior member of Britain's wartime enemy. 'I think I have heard about Peenemünde,' he said carefully.

'That's why the Russians brought me here, with my whole team from Nordhausen, to work with them on a bigger and better rocket,' said Helmut. 'We started by copying the V-2, but we are developing it much further.'

'I heard that hundreds of scientists were put onto trains by the Russians, back in 1946,' said Cocky.

'Maybe a hundred and seventy,' said Helmut. 'Plus families.'

'Did they all come here?'

'No, there are more in Podlipki, and in other workshops. We were in Podlipki for a couple of years, me and my wife Irmgard, who you've just met. We only came to this island in 1948.'

'An island, like Peenemünde?'

'Yes, the Russians even copied the idea of using a remote island.'

'Are we on a remote island?' asked Cocky. 'Where are we?'

'We're on Gorodomlya.' Helmut sounded affronted by Cocky's ignorance.

'I don't know why I'm here,' said Cocky. 'I really don't.'

'My workshop manager here, Preikschat, asked our main laboratory in Podlipki, near Moscow, for a guidance specialist,' explained Helmut, 'after their last visit to us here on the island. It seems that you're the only specialist anyone there could think of. I wanted another one, a particular Russian.[4] You've never worked on rockets before?'

'No,' admitted Cocky miserably.

'Come on. You were with Dr Dieminger at the Fraunhofer Institute? Working on radio-controlled direction finders?'

'No. Yes. Not worked with him, no,' said Cocky.

'Your British boss, Edward Appleton, worked with Dieminger, then?'

'Edward Appleton worked with Dieminger, I believe so. But not me, not directly.'

Helmut ploughed on, the tension showing on his face. 'Alright. What do you know about radio-controlled guidance systems? High frequency, wireless communications – anything?'

'Not much,' said Cocky.

Helmut took a deep breath, which scarcely concealed his anger and frustration. Despite looking and sounding exhausted, he explained how he and four of his 'best engineers' had grappled with radio-controlled guidance for the various rockets they had worked on. However good the system looked on paper, in practice it would cut out with interference or even a cloudy sky. In addition, the noise of the engine prevented the onboard antenna from receiving directional pulses transmitted from the ground.

'Every time we send up a miniature from our test stand, the damned thing goes the wrong way,' he said. Helmut mimed a rocket being launched, only for it to change direction in mid-air and come down a long way from its target. He explained how the shortages of components, the fact they had to manufacture their own equipment, and how the isolation of the group of scientists on Gorodomlya – who weren't party to developments in any of the other research institutes – contributed to an almost hopeless situation. 'Their policy is to keep us isolated,' he told Cocky. 'They seem to think we can magic solutions out of thin air.'

Cocky wondered whether German and even some Russian scientists were deliberately slowing the development of guidance systems. That strategy couldn't stop the Soviets from getting there in the end. But, if the moment of highest political tension could be allowed to pass, then a weapons scientist who slowed developments could also be an agent of peace. Cocky quietly resolved, as he sat on the Gröttrups' settee, to be an agent of peace.

Helmut was still talking about his technical development difficulties. 'That's why we asked for a guidance specialist. They sent you. They said you were definitely the right man. You're a guidance specialist, aren't you?'

'I'm an engineer,' said Cocky, 'with experience in electronics, and, yes, I know a little about the use of radio waves in guidance systems, but I'm not …'

Helmut looked crestfallen. He left the room.

'Do you like sausage?' asked the young girl on an errand from her mother in the kitchen. She was trying out her English and had been waiting for a moment to speak.

'Yes, thank you, I do like sausage,' said Cocky, smiling at her in relief.

'You had better be a guidance specialist,' said Helmut, putting his head back round the door. 'Or I don't know why we need you, and you might as well go back.'

'He can't go back, he's only just arrived,' said Irmgard lightly.

Helmut looked stumped.

'Leave him alone, Helmut, let him eat. You can talk about work tomorrow,' pleaded Irmgard.

'Electronics engineer. Huh.' Helmut stomped loudly up the wooden stairs.

Cocky ate his meal with Irmgard and the children, who skipped around, asking him questions. Dimly, he realised that Helmut was his new boss and that the interview hadn't gone well. The two children soon took themselves off to bed. Irmgard showed him to the top of the house, to a comfortable bedroom with its own tiny bathroom and even smaller anteroom, which formed a sort of apartment. Cocky was as pleased as he could be with his luxurious prison cell.

'Breakfast is at seven,' said Irmgard. 'Don't mind Helmut, he is just grumpy. Mr Grumps,' she said irreverently, and gave Cocky the same wide smile she had shown before.

Considering his incarcerated state, Cocky slept extremely well, and was up and about well before seven. He used his little bathroom and hung his few clothes in the small wardrobe. He breathed in the detergent smell of his clothes, reminding him of his mother's new flat in Bristol, where she had mastered her new twin tub for the first time.

Irmgard had told him she would wash his clothes along with her family's, but said she didn't like the tarry smell of the Russian washing soap. 'We get used to it, but I use as little as possible.'

'When I arrived last night, the steamer took me across a wide lake,' said Cocky at breakfast. 'What's it called?'

'I'm not sure,' hedged Helmut.

'Lake Seliger,' said Irmgard, 'in Tver Oblast.'

'The island we're on, what's it called?' asked Cocky.

'Gorodomlya,' said Irmgard.

'You mentioned the V-2 rocket last night,' said Cocky to Helmut.

'We've gone way past the V-2,' snapped Helmut.

Cocky felt like he had just been slapped.

Helmut's temper was boiling over. His English was ragged and accompanied by angry, jabbing finger gestures. 'It's 1951, nearly 1952, and we are way past the V-bloody-2. The Soviets are now working on their R-7, so that's five versions on from the V-2, each one getting bigger and longer range, each one needing a more accurate guidance system, because every bit of extra weight gets magnified by speed, so the bigger they are, the more scope there is for them to go totally astray.' Helmut was shouting but he sounded despairing. His reputation was resting on making sure these rockets could fly to the places they were supposed to go. His project, to design the next generation of rocket, which would travel further, carry a larger warhead, and land accurately on target, was starved of the co-operation and collaboration he needed from the Russians.

'Show me what you've got so far,' said Cocky. 'When we get to the workshop'.

It was only a ten-minute walk to the workshop but by the time they arrived, Cocky knew his best strategy would be to appear to co-operate with Helmut to develop an accurate guidance system for the Russian's R-7 rocket. He must not tip Helmut into any sort of rage. He had no idea how anyone could have assumed he was a rocket-guidance specialist. He could only imagine that the person or people who had put him forward were under intense pressure to name someone who could do it, or probably, more to the point, someone they could blame if he was unable to do it.

'Herr Gröttrup,' shouted someone as they arrived. 'Come and see these results.'

'Preikschat,'[5] acknowledged Helmut, without bothering to introduce him.

Cocky noticed that the team here were using experimental 'RM'[6] rockets like the ones at the Fraunhofer Institute. Their small size saved on materials, and allowed for different fuels and different casings, welds, nuts and bolts to be tried out, as well as different configurations for the electronics. There was a small test stand outside, from which they fired them regularly. The printout of results from the oscillometer was crude and inky, but it was enough to show that a particular configuration of components was moving the rocket towards better control of its lateral flight.

The morning was spent unpacking the experiments that had led to the data. At one point, Cocky realised Helmut had wandered off to another part of the workshop. He scribbled notes that would help him get up to speed with the rocket's capabilities and the particular challenges it presented. Rapidly, he lost his burning indignation about the uses to which a long-range rocket might be put.[7] He was surrounded by people who saw a successful rocket design as their only way home, back to freedom in Germany. Even so, he remained committed to making no difference to the Soviet's rocket, to keeping his head down and contributing to delays.

It was difficult to know who to trust. Cocky decided not to trust anyone at all. Even Irmgard. He wasn't sure where Helmut's true loyalties lay. He would never risk trying to uncover them. Helmut didn't introduce him around, and he decided that remaining incognito was the safest way. Russian workers mistook him for a German, whilst Germans assumed he was a newly arrived Russian.

Over lunch, when everyone else was standing around a makeshift canteen, which was no more than a trolley with an urn balanced on top of it, Cocky took the opportunity to pick up a few resistors and capacitors, and a length of wire. He pocketed them, resolving to pick up something else every day until he had the components he needed to build a radio set. It would give him a plan, a hope of escape, and a way to make contact with the outside world. He fingered the precious components in his pocket. To calm himself, he hummed 'Three little maids from school'.

To his horror, Helmut's Russian female security guard was checking everyone's pockets on the way out of the workshop at the end of the day. Cocky had time to pull the things out and stuff them back onto the benches, but not before one or two people had spotted his attempt. He was waved through to the exit doors, but it had been a close shave, and Cocky didn't know whether or not the people who had seen him with the radio components would tell Helmut. He was shaking with the stress of near discovery. He had to think of another, less obvious way to smuggle the components. He walked down to the lake's shore to try to get his bearings, and to recover. On the water's edge, a thin young man called Walter stood close to him and pointed out the sights and landmarks.

'This is Lake Seliger,' said Walter. 'We're in Tver Oblast, about 350km north-west of Moscow.'

Cocky started. He remembered the white pin in the huge map on the table at Obernkirchen. He had placed it in the water of this lake, approximately where he was now standing on the shore.

'And the island is Gorodomlya Island,' said Walter, 'in case you should ever need to know. Its nearest town on the mainland is Ostashkov.'

Cocky looked at Walter.

Walter stared straight ahead. 'There's a monastery called Nilov, on that island over there, Stolobny Island. You can just about see it from here,' he said, sounding like a tourist.

Cocky tried to memorise the place names. 'It's a beautiful lake,' he responded, sounding like a tourist.

'I have those components you left on the bench,' said Walter. 'I'll leave them in a tin behind a rock – the porpoise-shaped rock you can see over there by the treeline at the back of the beach.'

'How did you manage to …?' began Cocky, but Walter had gone, and was making his way through the woods back to his home.

Helmut appeared by Cocky's side.

'Walter's told me where we are,' said Cocky, realising that Helmut must have seen the brief exchange between him and Walter. 'Lake Seliger.'

'Yes, that's the name of the lake,' said Helmut. 'Doesn't mean that you know where we are, though – we could be absolutely anywhere in Russia. There are lakes and there are islands everywhere. Monasteries, too.'

'I suppose that's true,' said Cocky.

But that evening in his room, he recreated from memory the map of Lake Seliger and Gorodomlya Island in its context in Tver Oblast, halfway between Moscow and Leningrad. He would add in more detail as he became aware of it, but for now, he knew he was exactly where he had pressed in the white pin, and he also knew from that day at the Unit with Geoff's map that Gorodomlya Island was an outpost, or branch, of Korolev's main workshop, N11-88.

He went for a walk on the beach the same evening, and casually picked up Walter's box from behind the porpoise-shaped rock, carefully buried in the sand. The resistors and capacitors were all there, as well as a vacuum tube and a length of copper wire. Even a small pair of wire strippers, which Walter had thoughtfully added to the stash. Cocky started planning what he would need to pick up tomorrow. A variable tuning capacitor might be too much to hope for. He could add the components onto the weekly supplies list for the workshop, and hope that by listing them in with lots of other bits and pieces, his requests would not be noticed. He didn't know why Walter was willing to do the smuggling of components from the workshop, but for now, it was enough that he had an accomplice. At this rate, he might be able to build a radio by the end of the month. He was desperate to talk to Eric. A radio was the only form of communication open to him.

He went downstairs to find Irmgard, looking for a little of the friendliness of the day before. Irmgard and Helmut were arguing, and it sounded a bit heated, so Cocky went back upstairs. He was just about to brave another descent to the kitchen, when Irmgard knocked on his door and passed him a tray set with food on a plate, some flat beer, and a big hunk of fresh bread.

'The bread is a rare treat,' she said. 'Don't waste it.'

Cocky thought he would be eating with the family, but that wasn't going to be the pattern. He guessed that Helmut didn't want him sharing his domestic space as well as his workshop. He contented himself with the good food, the comfortable place, and with plans for his radio. He looked through his technical notes from the day and thought about tomorrow. There was nothing to do but have an early night.

Over the next weeks, Irmgard managed to tell Cocky that there was a social life of sorts on the island, which had grown from the beginnings of the transplanted community in 1946. There was a clubhouse where plays and concerts were held and people celebrated Christmas and New Year. There were, or had been, parties and reading evenings, where people shared their favourites. There were shopping trips under guard for the housewives, and picnics on the beach. The local market in Ostashkov had supplies, although it was very seasonal and there were gluts of one vegetable that everyone would tire of eating before a glut of something else would take its place. A school at the club was run by a German professor, with some Russian teachers. There was also a kindergarten, and a clinic with a young doctor who had been a medical student before the war. He drew the line at midwifery, though, and the few babies born in the past five years had been delivered by a female dentist, with help from a local Russian. Irmgard had assisted at a few births.

N11-88, Branch 1, 1952

My father worked here for around fifteen to seventeen months, from the end of 1951, through 1952 and into the spring of 1953. He recalls Werner Albring and Karl Preikschat leaving the island in June 1952 with their families. He remembers Kurt Magnus being around. Until recently, a large group of radio control specialists had been using the workshop at Gorodomlya, which was set up for that purpose. The workshop had produced all of the equipment for radio control of the Russian A4 rocket, which was the direct copy of the V-2.

By the time my father was there, most of the feverish work on the rocket called the G-4, or R14, which turned out to be the last main project of the workshop, had been passed to the Soviets. Irmgard writes that Helmut Gröttrup presented the R-14 design and technology, and the Soviets kept on coming back to him for six months, wanting more details and explanations.[8]

The fixation on 'high frequency' radio control was almost total, despite the fact that in the summer of 1952, the American MIT[9] laboratory was coming up with inertial, computational-based solutions. The Russians must have heard of these developments, because they tasked Helmut with developing an electronic computer in the last few months of his time on the island.

One midday in summer 1952,[10] a wave of tension swept around the workshop, because Korolev and some other officials were coming over from N11-88 laboratory in Podlipki, near Moscow, in the following weeks, to assess Gröttrup's progress with the rocket design work. Helmut went into overdrive, issuing orders and countermanding them, and hauling people to one side to assess the progress of their particular project area.

Just before the end of the working day, it was Cocky's turn to be grilled. Helmut often veered between friendliness and threat, and this time Cocky wondered if he was about to be sacked. Sacking would be alright if it meant he could go home, but Helmut didn't have the authority to send him home.

'You've been here nearly six months,' said Helmut. 'You were brought here to bring our work on guidance systems to fruition. You are supposed to be a specialist! They already have a Guidance workshop, N11-885 – run by Ryazansky and Pilyugin, and another one, N11-10, run by Kuznetsov, focusing on just the gyroscopic platform.'

'The gyroscopic platform?' said Cocky.

'Yes, yes, and Kuznetsov is one of their best. These Russians are taking over our work.'

'It looks as if Kuznetsov is moving towards an inertial system,' said Cocky.

During his time in the workshop, Cocky had noticed how radio-controlled guidance systems were the only form of guidance being investigated. So far, he had worked on improvements to that system, without mentioning alternatives. Now he realised that Kuznetsov must be researching an inertial system, because of the focus on gyroscopes.[11]

'The original V-2 had gyroscopes,' said Helmut.

'Yes, for stabilising flight. Not guidance,' affirmed Cocky.

Helmut suddenly started listening.

Cocky outlined the system he had been aware of since 1950, when Eric had shown him a very early version of one he had salvaged from V-2 scrap parts. He explained to Helmut the fundamentals of how it worked when connected up to an onboard analogue computer.

'The gyros, three of them, plus accelerometer, determine the exact position and velocity of the missile as it flies, comparing it with the preset flight path. Electronic signals transmit through to servomotors, which move the control surfaces, so the rocket body is kept on course. It doesn't rely on radio control from operators on the ground – it's different altogether.'

'Must be something like the one Kuznetsov is looking at,' mused Helmut. 'I've only heard rumours, though. As usual, I can't get hold of the detail.'

Cocky wanted to get into a position where he had some control over development time. He didn't want Kuznetsov racing ahead, so he inserted himself into the equation.

'No need for control towers, antennas and the like?' Helmut wondered.

'No need for them,' confirmed Cocky. 'It automatically calculates the difference between programmed and actual trajectory. Adjusting flight surfaces to keep on course. Like a pigeon.'

'We don't have time for your British jokes,' dismissed Helmut. 'I know about your Crazy People, your Goons, laughing at the war, laughing at everything.'

'It isn't a joke,' said Cocky. 'Pigeons really do ...'

'Could this method work better than the radio guidance we've been working on?' cut in Helmut, his voice soft now. 'For the larger rocket designs?'

'It would need to integrate with the rest of the rocket body,' said Cocky, knowing this was difficult because the rest of the rocket body was being designed elsewhere.

There was a silence. Helmut offered Cocky a rare and precious cigarette. Cocky declined it, but he picked up on the abrupt change in Helmut's attitude towards him.

'You can build one to show Korolev – by next visit?' asked Helmut.

'I will need a basic telemetry unit, the use of the analogue computer we've got in the workshop, a set of at least three gyroscopes, two lateral accelerometers, a battery, some decent speakers, and a pair of brown leather headphones,' said Cocky, who was thinking about his own radio.

'Whatever you need,' said Helmut. 'I will make sure it gets to you by next week.'[12]

Helmut's abrupt swing to unexpected friendliness was welcome on one level, but it brought its own difficulties, because he started inviting Cocky to share family meals, just as he wanted to be on his own with his radio, trying to make contact with Eric and others at the ASRU.

'As long as I can be in bed by nine,' he would say to Helmut, citing the demonstration of the onboard guidance system as the reason he needed to get a good night's sleep. He also claimed his own small team of technicians, and his own area at the back of the workshop.

'Of course, of course,' said Helmut. 'Whatever you need.'

Cocky knew he couldn't build one in a few weeks. He didn't know if he could build one at all. But he knew someone who could find out more about it – Eric, who had been thinking about inertial guidance in his study at Obernkirchen since at least 1950. Rumours circulating at the ASRU also intimated that Eric was in touch with the Americans, who had German rocket scientists Wernher von Braun and Helmut Hölzer helping to develop their own rockets.

Eric, Cocky felt sure, would have emerging information about the Americans' fledgling inertial guidance systems. But how could he expect Eric to risk giving away knowledge about it that would surely help the Russians move faster towards an inertial system of their own? He would be giving Eric a terrible dilemma. Cocky thought this through and hoped that between them, he and Eric could come up with a compromise. They weren't going to genuinely assist the Russians. Since the Fraunhofer mission, he and Eric had often practised how to appear to be giving full technical information. But by missing out a crucial fact, the information could be rendered useless. His head was reeling with the stress of it all.

Secret Radios

Cocky's homemade radio set was nearly finished, constructed from components that either he or Walter had appropriated from the workshop. He had speakers, but he needed headphones to make sure that Helmut wouldn't be alerted to the existence of the radio in the house. Materials to hold the speakers against the ears were just not available. It was frustrating, because he was keen to communicate with the Signals Unit, and with Eric. He had tried tying a towel around his head, or a pair of socks, without success. His need for headphones started to consume him. Eventually, he took the risk of asking Irmgard if she could get him some headphones on the housewives' weekly trip to Ostashkov market. She laughed, reminding Cocky that they were in Russia, not Germany. Then, Irmgard had asked what he needed them for. He was ready for this question, aware that if Irmgard reported the existence of his radio to Helmut, things could get very difficult.

'I've made a small radio, which I use to listen to classical music,' said Cocky. 'But I can't play it loud enough to enjoy, without some headphones.'

Irmgard thought for a while about the question of headphones. Then she offered Cocky a pair of her own earmuffs from the winter of 1946, their first winter in Russia. 'We didn't have hot water at all, had to heat everything up on a creaky old stove. The earmuffs are bright red, and furry, but you can make them into headphones if you want.'

'No one's going to see me wearing them, so the colour doesn't matter,' he told her. 'I'll be able to stitch in a couple of speakers.'

'I'm happy to do some stitching for you,' smiled Irmgard, 'in case that's not your best skill.'

A few evenings and some surreptitious stitching later, and the headphones were ready to use. Cocky spent the day in a state of nervous excitement, knowing that tonight might be the moment when he would connect again with the Unit in Obernkirchen, and with Eric in particular. He and Eric had worked out their own code, which Cocky would use now. He wrote this out on paper, memorised it and then burned the paper before switching on his radio set. He ensured it was in 'transmit' mode, and moved the pointer along the wire coil, gently with a matchstick, looking for the right frequency.

The faded earmuffs gave him a comical look, but there were no witnesses, as Cocky tried for over an hour to find the frequency he was looking for. He sent out a few experimental transmissions, hoping for a response, but no one came back with a signal. Cocky eventually switched off the set, stowed it in the wardrobe, and went to bed. He remembered that Eric himself hardly ever did the actual interception of messages, and that no one else would understand his code. The operators at the ASRU would glide over it, looking for something more knowable, on one of the Russians' known frequencies, not the specific one he was using. He forced himself to sleep, intending to try again tomorrow.

However, on the next evening, there was a party at the club for someone's birthday. It would have been odd if Cocky hadn't gone, so he went along, resolving to contact Eric at the next opportunity. Partway through the evening, he realised he hadn't seen Helmut for about an hour, and he had an attack of anxiety about him discovering the radio. Helmut had never climbed the stairs to Cocky's topmost rooms, as far as he knew, but there was nothing to stop him doing so. He moved through the clubhouse, scanning each room. He only relaxed when he spotted Helmut by the buffet table, deep in conversation with Walter. Even so, he would have to work harder on concealing the radio apparatus. He would need to be vigilant about hiding it every time he left the room. He was getting better at the art of deception, but he still had glaring lapses, and he kicked himself that the radio was in an obvious place, the wardrobe. He left the party early.

The toilet cistern was the hiding place he had in mind, but it was too wet in there, and the cistern lid was heavy and likely to crack if he dropped it. Behind the wardrobe, another good place, but it was too near the wall, with no nooks or crannies. Moving the heavy wardrobe forward proved impossible.

The floorboards – he rolled up the rug and tested each one for looseness. They were all sturdily nailed down.

His laundry bag. The one he collected dirty clothing in for the weekly wash. He kept it in the bottom of the wardrobe, and took it downstairs himself on Saturday mornings. It was probably the least appealing place for anyone else to root around in, and he carefully placed the collection of wires and coils soldered to a piece of Bakelite into the cloth bag, surrounding it with dirty socks.

A few evenings later, Cocky pulled the apparatus out of its smelly hiding place, jammed a chair up against the door, spooled the antenna out as far as he could, and switched it on. He had been experimenting with locating the frequencies he knew the Unit focused on, and his aim this evening was simply to let them know that he was alive and transmitting. A blast of classical music came out at him from the set, so he changed the mode from 'receive' to 'transmit' and he started to move the matchstick lever along the surface of the wired coil. With the mode now on 'alternating transmit/receive', there were ten minutes of static and indecipherable bits of signal and speech. Then, Cocky suddenly heard a signal that he recognised. It was just a pattern of beeps, dits and dots, but it was clear, and the longer he listened to it, the more sure he was that this was the call sign of an operator at the ASRU. With the lever on the coil, he pressed out an identical pattern, and waited for it to repeat. It came back again, slightly more slowly, and he repeated it back at the same tempo. The pattern came again, even slower, and Cocky matched the slowness exactly.

The operator moved to standard Morse code, and asked, 'Identity?'

Cocky tapped in his last Signals Unit identity code.

The operator repeated Cocky's code, and Cocky confirmed it.

'Location?' came the operator's signal.

'Russia, Lake Seliger, between Leningrad and Moscow,' tapped Cocky, using the code he had practised and memorised. Then he signalled 'WHITE PIN'.

There was a significant pause. Cocky wondered if the operator was asking other operators in the listening room, 'Does anyone know what "White Pin" could possibly mean?'

'Ten p.m. tomorrow night,' came the message back from the operator.

Cocky tapped in Eric's call sign, together with the signal for a question. 'Eric?'

He wasn't sure if the operator received it. The radio had moved away from the frequency – either the signal had been jammed or some interference had intervened. Cocky tried for a bit longer to resume the call. He had to eventually stop trying, switch it off, wrap it inside the laundry bag, and try to sleep. 'Ten o'clock tomorrow night,' he said to himself, over and over, as if he was in any danger of forgetting such a momentous date.

The next day at the workshop passed in a blur of excitement for Cocky, who was collecting in his mind all the information and questions he had to ask Eric. He was wondering too if Eric could exert pressure on the British government to secure his release.

Irmgard was surprised to hear that Cocky was joining them for the evening meal. She made an extra special effort, and the children were delightfully well behaved. But the biggest surprise was Helmut, congenially offering light conversation and anecdotes, jokes and vodka. It was seductive, but Cocky kept his wits about him, joining in only as far as politeness allowed. He resolutely declined the vodka.

At nine o'clock, Cocky went upstairs at the same time as the children went to bed. He assembled the radio set from the laundry bag, and spent a few tense minutes tuning into the right frequency for Eric at the ASRU. He prepared a few signal messages, in his and Eric's own code, and waited jumpily for 10 p.m.

At ten o'clock on the dot, Eric's call sign came through. Cocky replied immediately with his own signal. There was no cheering, no hurrahs or gasps of relief, but Cocky's heart was pounding with excitement and he sensed that Eric must be feeling the same.

'Location?' asked Eric.

Cocky gave his exact location.

'Are you alright?' came Eric's first question.

'Yes,' responded Cocky. 'I am in a Russian camp for rocket scientists. I have to design a gyroscopic guidance system. Can you help?'

There was a long pause.

'I can help. What do you need?' Eric's response finally came through.

'I need to know everything. I don't know enough about it,' tapped Cocky.

'Each of three gyroscopes, when wired up to an onboard control unit, sense deviation from the desired flight path for each of pitch, yaw and roll,' signalled Eric. 'The accelerometer integrates this with velocity data. Hölzer has improved his electronic numerical integrator. His analogue computer.'

'Tell me about Hölzer's work,' said Cocky. 'Just the bare minimum, to protect me from …'

'Fine weather at present,' affirmed Eric.

'I am staying out of the sun,' replied Cocky.

'I am pleased you are enjoying your holiday,' tapped out Eric.

'I can't stay much longer,' tapped Cocky. 'It's too hot.'

'Full itinerary transmission tomorrow night, 10 p.m.,' affirmed Eric, using their own code.

'Over and out,' responded Cocky.

'Over and out.'

Tasks and Inspections

The next day at the laboratory, a shimmer of tension went around the workbenches, as news spread about the next 'Commission inspection' from Korolev and his colleagues. Helmut came over to Cocky with a printed list that had come through on the ancient teleprinter. He stood with the list in his hand, deciphering the instructions and informing his team at the same time.

Cocky wasn't usually involved in visits or inspections from the main workshop in Podlipki. He kept a low profile, practically hiding in corners of the building, mentally trying to absorb any information that could be passed on to Eric. This inspection visit took place in October 1952, and was a very different matter for him.

The scientist who Helmut had repeatedly asked to have working with him, V.I. Kuznetsov, was finally attending a meeting with Gröttrup to discuss the challenges of accurate guidance. The radio-controlled guidance efforts had been officially discontinued on Gorodomlya, but there remained an urgent need to develop alternatives. Helmut had warned Cocky that Viktor Kuznetsov was coming as part of the delegation, and that he would be interested in the work on the gyroscopic guidance system.

'You must share your work with Kuznetsov,' Helmut told Cocky.

'I'm still experimenting with positioning the contact points for the wiring from the gimbal,' Cocky replied.

'Yes, well, Kuznetsov will be able to help you with that. Work together,' urged Helmut. 'And I want you to choose a few technicians to demonstrate it with you, including Russians.'

Cocky knew the only technician he wanted to work with was Walter.

***

This may have been the invention that my father, as an electronics engineer, envisaged as a potential contribution to the Russian rocket. He is clear that he worked against his masters – trying not to communicate useful expertise. He and Walter colluded to go down dead-ends. They worked slowly, ordering components that were difficult to get, and introducing errors into their work.

Transmitting to the British and Americans

Cocky hardly slept that night because of the sheer excitement of having made radio contact with Eric, but also because Irmgard and Helmut were arguing loudly. Below Cocky's top-floor rooms, snatches of barely suppressed anger reached him. Helmut snarled his words and Irmgard tried to maintain a semblance of calm.

Their arguments were often about continuing to live in Russia, which Helmut was more positive about than Irmgard. His work on Gorodomlya was frustrating and he felt sidelined, probably because he actually had been sidelined. Branch 1 was definitely subordinate to the main institute in Moscow. He wanted to get back to more prestigious and interesting work, but the Moscow scientists had determined to reduce their dependency on German expertise. Helmut often mentioned the irony of Gorodomlya being a 'branch' of Korolev's main laboratory. He would shout: 'You know what happens to a branch when there's no more fruit? It gets lopped off!'

Irmgard wanted only to go back to Germany. She had proved herself good at adapting to both Moscow and Gorodomlya. She was a community leader and a strong personality. But she wanted a more normal upbringing for their children. She was sometimes scathing about Helmut's all-consuming work with the rockets, and their marriage at this time wasn't happy. 'Remember, we agreed we would stay in Russia for the five years, then go back to Germany. The children – I want them to go to school in Germany, to live in freedom.'

'We don't have any choice about staying in Russia,' Helmut would counter. 'Those first couple of years in Russia, when we were living in Podlipki, and I was going out to the Russian test ground at Kasputin Yar, helping Korolev with the test rig and the early launch experiments, I really thought …'

'I always thought', interrupted Irmgard, 'you would help get their rocket to a certain point, and then they'd let us go home. I still can't believe they've sent us here, to this backwater.[13] Quite literally, we're not part of developments anymore.'

'There is nothing for us either in Germany,' Helmut would shout. 'No work for me, not in my field. This is our best chance, for the future of our children – we could live in Podlipki, or even an apartment in Moscow, we'd have a dacha in the countryside …'

'No,' Irmgard would shout back. 'They will use you like they did at first. They will take your expertise, tolerate you as long as it suits them, and all the while they will be training up Russian scientists to do your work. They will not reward you in any way, because they do not want the world to know their precious rocket was designed by Germans!'[14]

* * *

My creative deduction as to why Finnish classical musicians were invited to the island by the Russians on the request of Helmut Gröttrup is that he was, by this stage in his career on the island, desperate to demonstrate cultural equality with his Russian masters. Helmut often said, 'Classical music is the route to personal connection with the Russians.'

His request may have been agreed to by the Russians, specifically Korolev and Ustinov, out of their sense of guilt that they had worked with Gröttrup as equals at first, then systematically sidelined him whilst using his designs for their R14 rocket.

Helmut had spoken with Dmitry Ustinov, the head of the Soviet Ministry of Armaments, in 1949. At that time, Helmut was Chief Engineer on the island, and was expected to talk to the important visitor. Ustinov showed interest in the Germans' cultural activities, offering them a trip to Moscow to see a concert or play. The trip never took place.

Irmgard usually organised and attended the community's homespun concerts on her own, without input from Helmut, who was always too busy.

'I love those concerts,' said Irmgard. 'Even though they've played the same set of pieces now for nearly five years, each time they mean more. Bach, Strauss, Handel ... But Helmut, you've never wanted to come with me.'

'I like classical music, dammit woman!' shouted Helmut.

'Keep your voice down, dear,' said Irmgard. 'The children ... and Cocky ...'

Their voices sank lower, out of earshot, and Cocky finally managed to sleep.

Next day at the workshop, Walter came shambling over, his long legs and arms uncoordinated, but his mind razor-sharp.

'How's your radio?' asked Walter. 'Enjoying listening to music on it?'

'Very much,' said Cocky.

'I made one too,' said Walter. 'When I first arrived. I keep it in a shoebox under the bed.'

'Mine's tucked away, too,' said Cocky. 'Even though it's only a receiver, no transmitter.'

'Do you want to build in a transmitter?' asked Walter a bit later.

'No thanks,' said Cocky. 'Transmissions from here, if they got picked up, would look highly suspicious.'

'I thought I picked up something last night,' said Walter. 'A strong signal, almost on my own frequency. Just a shade away.'

'Oh?' said Cocky. This was clearly a warning.

Ferry connecting Gorodomlya Island with Ostashkov. (*Anatoly Zak, 2002*)

My father and me in Hasfield, Gloucestershire, about 1970. (*Author's collection*)

Interior of the workshop on Gorodomlya Island, 1952. (*Werner Albring*)

Barbed wire surrounding Gorodomlya Island, strands 20cm apart. (*Georgy Karlov*)

Bruce Neville Cox (Cocky) with family members, 1958. (*B.N. Cox*)

Eric Ackermann as a young man at the Telecommunications Research Establishment, Swanage, UK. (*David Haysom*)

RAF Wilmslow, where B.N. Cox did his basic training. It was closed in 1962. (*Forces War Records (RAF)*)

Bedford truck, 1939–45. (*Imperial War Museums*)

The London Blitz: damage caused by V-2 rocket attacks in Britain, 1945. (*Wikimedia Commons*)

V-2 rocket launch from Peenemünde, Germany, 1942.
(*Wikimedia Commons*)

In 1938, Korolev was sent to a gulag by Stalin in the 'Great Terror', for obstructing weapon development work. Paroled in 1944, this photograph shows a still malnourished Korolev in 1946.
(*Boris Chertok*)

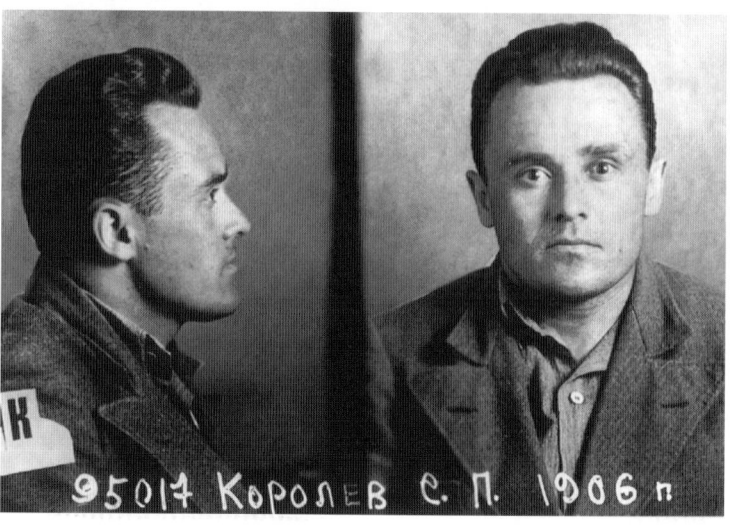

Sergei Pavlovich Korolev, mugshot taken by the NKVD (KGB), 1938.
(*Natalya Koroleva*)

The barrack block, RAF Changi, 1948. *(B.N. Cox)*

RAF Obernkirchen in 1958, looking towards the entrance. B.N. Cox says in 1951, the facilities were 'primitive'. *(G.M. Stewart)*

The Brocken tower in 1951, before the iconic red-and-white transmitter was constructed. *(R. Demuth)*

Helmut Gröttrup in 1958.
(*Ursula Gröttrup*)

Branch 1 workshop on Gorodomlya Island, 1952.
(*TsIIMash*)

Helmut and Irmgard Gröttrup, boat trip, 1950. (*Ursula Gröttrup*)

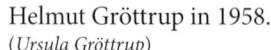

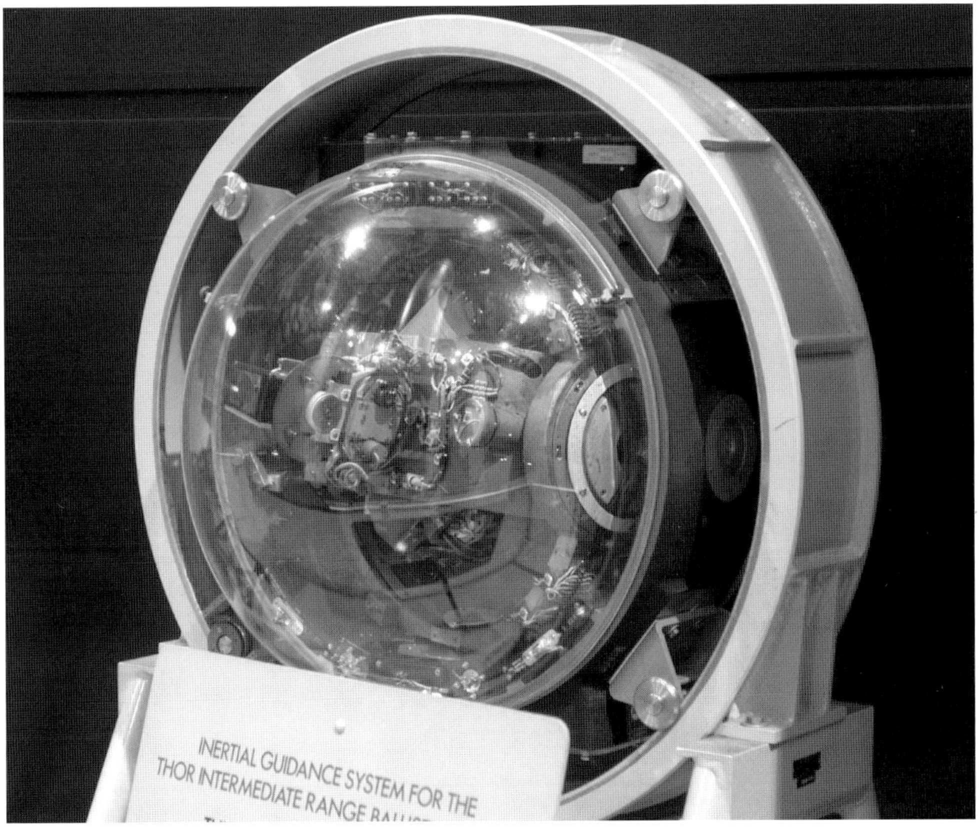

Inertial guidance system, produced 1959–62 for the US Thor IRBM (Intermediate Range Ballistic Missile), showing gyroscopes. (*US Air Force*)

Electronic inertial guidance system for project SPIRE. Developed by MIT from Hölzer's system in the early 1950s. (*Sanjay Acharya*)

Sketch of the G-4 (R-14) design concept by Konrad Toebe, who was a member of Helmut Gröttrup's team on Gorodomlya Island from 1946 to 1952.
(SchmiAlf, Wikimedia Commons)

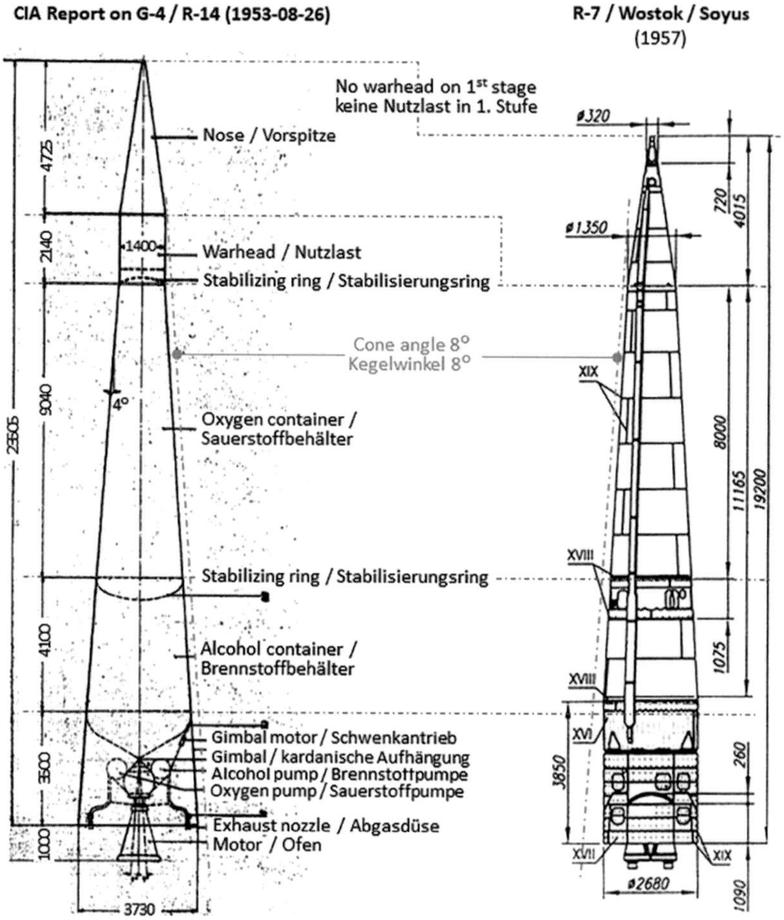

Zoom call in Covid lockdown, 2021. The author, brother David and father, B.N. Cox, aged 98.
(Author archive)

'I do some transmitting,' murmured Walter as their two heads bent over some wiring. 'There are only a couple of wavelengths that work long distance. I mean really long distance – the States.'

Cocky took in the implications of what Walter was telling him. He appeared to be admitting he was communicating directly with listeners in the US. He had been on Gorodomlya since 1946, so it was possible he was responsible for some of the Americans' knowledge about Soviet rocketry.

'If I ever do get a transmitter,' said Cocky, 'I'll move away from that frequency.'

'Good idea. These two wires should go there, where the contact will be most secure,' said Walter.

Helmut was hovering.

'What will you have to show by next week?' he queried. 'For this Kuznetsov and Minister Ustinov?'

'Not much,' said Cocky, before realising that was the wrong answer.

Helmut nearly exploded with the effort of containing his temper.

'We'll have a wired-up set of gyroscopes and a lateral accelerometer, with the wires leading to an analogue computer,' said Walter. 'By next week, the computer will be able to receive data from the movement of the gyroscope inside its gimbal, so the measurements between the contact points at the edge of the gimbal ...'

'Will it prove that we have cracked this new form of guidance system?' barked Helmut. 'Because that's all I care about. That is all they care about too. Have you got something that will keep the bloody rocket on course, or haven't you?'

'Yes. Yes we have,' said Cocky and Walter. 'We've got the basics.'

Skating, Tennis, Boat Trips and Concerts

When I was researching this episode, I was trying to understand the context of how my father eventually escaped from the island – in disguise, as part of a visiting Finnish quartet. That an island dedicated to creating rockets should have artistic, cultural and sporting facilities at all, let alone a visiting string quartet, seemed to me most strange.

I found many anecdotes demonstrating a lively cultural life in Irmgard Gröttrup's *Rocket Wife*, as well as Werner Albring's *Gorodomlya Island*. Albring's family was repatriated in June 1952, but he had been there since the start, in October 1946. Albring estimates the population at that time to be 400 Germans, including families, sometimes reaching 500, due to an influx of Russian workers.

Initially, the Soviets had expected the 'rocket specialists' to want to stay in Russia, and for the twenty 'top' families, including the Gröttrups, the first two years were spent in Podlipki, a suburb of Moscow. There were state-sponsored trips to the Bolshoi Ballet, concerts at the Tchaikovsky Hall, the theatre and opera. The Russians had shared their appreciation of music and their cultural institutions with these similarly cultured Germans.

On Gorodomlya, people found ways to create entertainment and practise their own culture. Albring says that 'New Year's Eve celebrations of 1947/48 invited a small circle of friends to our apartment for an "Olympic Feast of the Gods"', where everyone dressed up and pretended that the meagre fare they'd bought from the island shop was in fact a sumptuous feast. Guests had brought 'a literary contribution as a gift to the hosts'.

In February 1947, the islanders had celebrated Fasching carnival. Christmas and New Year were high spots, with families determined to keep traditions. On New Year's Eve in 1952, there was a party at the club, with 'Russian colleagues rehearsing a programme', and there was a chamber group, a small orchestra, a dance band, and a girls' choir.

Two German theatre companies formed from friendship groups, and they all learned parts and devised costumes. They created a 'theatre stage' in the woods for outdoor performances, and in the winter, performed at the club, without dressing rooms. Actors changed at home and walked, with a coat over their costume. Plays mentioned by Albring include Shakespeare's *As You Like It*, Molière's *Tartuffe*, and Klabund's *Der Kreidekreis*, or *The Chalk Circle*.

Irmgard Gröttrup was in a group of eight friends who felled trees, cleared space, made a form of cement, and constructed a tennis court. They enjoyed evening tennis matches for as long as the light lasted, stopping only in winter. This home-built tennis court was still in existence until recently.

Winters were long and harsh. From October to April, the snow and ice meant only winter sports were possible – such as skiing through the forest and down hillsides, and skating on the island's inland lake. A lot of energy went into chopping wood for heating and cooking, and guarded trips across the ice to Ostashkov meant dragging a sled for two hours. In summer, an annual boat trip was a highlight, and the steamer would fill up with people and picnics, and then make for any of the many other islands on the lake, which were mostly uninhabited. Swimming in the lake was another pastime, as well as sunbathing, hiking through the forest, and meeting up at friends' apartments for coffee, on the rare occasions when it was available. People had cats and dogs, and one man even kept some goats, bought from the market at Ostashkov. The Gröttrup family kept a tamed raven, and a young boy nursed a heron chick back to health by catching frogs from the reeds at the edge of the lake.

By 1952, much of the cultural activity had stopped. The first group of 'non-essential' Germans had been sent home. A vocal faction of islanders objected to the cultural activities because they didn't want the Russians to consider them settled. A nervous restlessness was always evident, because the vast majority of people wanted to return to Germany – the musical and theatrical activities helped pass the time, but weren't a reason to stay.

The question of why the Soviets agreed to invite the quartet is interesting. Korolev had been an equal-status colleague with Gröttrup at Nordhausen-Bleicherode, and in 1947, on Kapustin Yar launch test site, they had embraced[15] in the excitement of a successful rocket launch.

In his book, Werner Albring says of a Commission meeting on Gorodomlya, in October 1952:

> Herr Gröttrup told me that Mr Korolev had come to see him in his office once more to discuss our final project (the R-14 rocket). According to Gröttrup, Korolev seemed depressed and acted as if they were seeing each other for the last time.

This may have been the meeting where Gröttrup asked permission for a quartet to visit the island for the pre-Christmas concert. Gröttrup had been demoted from Chief Design Engineer in 1950, although he continued with the role because his successor, Herr Hoch, died from appendicitis four days later. Korolev may have granted Gröttrup's request for a concert by a visiting string quartet, as consolation or apology for the way Gröttrup had received no recognition or reward. The next time the Commission visited the island, Korolev did not attend.

Finnish Christmas Concert

'There's a delegation from Moscow next week,' Helmut said at teatime. 'There's a rumour they want to close the facility here. We must demonstrate progress, especially with the guidance system. I want some evening entertainment to offer our visitors. We can get to know Ustinov and Korolev, impress them with our understanding of high culture. The settlement orchestra can put on a concert.'

Irmgard blanched. The small orchestra was a random array of string players, brass players and percussionists, with only the sheet music that people had brought with them in 1946. Occasionally, this music was performed, but these recitals were for the rest of the community. There had never been the capability to put on a concert for outsiders, and many musicians had already left the island.

Cocky caught the end of the conversation about the concert. He suggested that he design a poster for it, and Helmut immediately accepted.

'Classical music is the route to achieving personal connection with the Russians,' stated Helmut. 'They love it, particularly Russian composers.'

'What about making a request for professional musicians?' suggested Irmgard tactfully.

'Yes. Good idea,' said Helmut. 'We must get permission to invite some professional musicians. A small group perhaps – not a whole orchestra.'

'A string quartet?' suggested Cocky.

'Give me the name of a suitable professional quartet', said Helmut, 'and I will make a request for them to perform here.'

'Maybe a Russian quartet would be allowed in, just for the evening,' said Irmgard.

'I don't know a Russian one,' said Cocky. 'The Russians go for full orchestras.'

Later, Irmgard made her way up the stairs to Cocky's rooms to discuss it further. She never came across the threshold, only ever standing in the doorway. Usually, it was a rushed, almost guilty conversation. On this occasion, though, Helmut had sent her.

'Helmut needs the name of a professional quartet,' she said. 'He is prepared to ask Moscow. If they agree, and we tell them who to invite, they will organise travel, get the visas and so on. We can't make direct contact; all our communication has to go through them.'

'What about a Finnish quartet?' said Cocky, who had been mulling over the escape opportunity presented by non-Russians. 'I know of several of those, but no idea of names. I'll have to listen to my radio.'

'Just the names will do at this stage,' said Irmgard. 'A Finnish quartet might be acceptable, if they don't mind coming here to play.'

'I'll be down in half an hour,' said Cocky. 'I've no idea about Finnish spelling, but I'll give it a try.'

Irmgard disappeared, and Cocky plugged into a Finnish radio station he knew, the one that transmitted the clearest classical music. He twiddled the matchstick, and heard a young man's voice introducing a piece, which Cocky realised by the cadence of his speech, must be his own composition.

It was Einojuhani Rautavaara, introducing his String Quartet No. 1, *Quartettino*. Cocky listened all the way through, about fifteen minutes. It was exciting, quirky music, modern, but he didn't have leisure to enjoy it. He listened out for the name of the performers, and wrote down what he heard. The group's name sounded like 'Strava',[16] and they were from Helsinki. He went downstairs with his approximations of their names on a piece of paper.

'Finnish?' said Helmut. 'This won't get approved.'

'I can't get Russian music on my set,' said Cocky. This wasn't true, but Cocky could see no value in inviting Russian musicians. He wanted non-Russians, people not under Kremlin orders, new people from a non-Soviet country.

'This Finnish quartet is really good, and they're young,' said Cocky.

'All men?' asked Helmut.

'One is female,' replied Cocky, hoping that his understanding of their names was correct.

'Good, good,' said Helmut with interest. 'It will be nice to see new faces.'

Irmgard looked hurt. 'They will be here for just one evening,' she reminded him sternly. 'They will play in the concert, then travel home the next day.'

'If they come,' said Helmut. 'They may not be available. They may not want to travel. Finns may not wish to play for Russians. We are a ten-hour drive from Helsinki.'

'They will need to stay overnight,' said Irmgard. 'In a hotel in Ostashkov, perhaps. Or I could ask amongst the housewives and see who has room.'

'Wait and see if we can get them to come at all,' said Helmut. 'You never know what will be allowed and what not. This is the first time we've asked for anything cultural to come to the island.'

'It's the first time we've asked for anything at all, since the saga of the horses,' said Irmgard.

'The horses were an escape threat,' said Helmut. 'Of course they were refused.'

'A quartet coming in from Finland could be an escape threat,' said Irmgard. 'They will guard them like they do us women on a shopping trip. Four in, four out.'

'Russians love classical music,' said Helmut, 'second only to Russian literature.'

'It's certain to impress our friends in Moscow,' smiled Irmgard perceptively.

Chapter 5

Escape from Gorodomlya, Spring 1953

Walter

Cocky became more and more focused on the idea of escape from Gorodomlya. He knew Lake Seliger was a day's drive from the Finnish border. The highly guarded compound, the remote island location, and the fact he had no passport were seemingly insurmountable barriers. The biggest fear he had, though, was his knowledge that if an escape attempt failed, he would be disposed of, and his 'disappearance' would be explained away as an unfortunate accident or illness, causing hardly a ripple. He knew of several scientists who had given freely of their knowledge, but instead of being recognised for their contribution, were never heard of again. The 'disappeared' were occasionally referred to with a shudder, but no one ever appeared to think this could be their own fate.

Cocky was on to the testing phase of his gyroscopic platform. He had grappled with finding a way to simulate the buffeting and random forces of the lower atmosphere, so that his triangular arrangement of gyroscopes, accelerometer and analogue computer could receive signals from the rocket's movement. To test his prototype, he needed to simulate atmospheric turbulence, to mock up the action of radiation flares and random weather events. He thought about the miniature rockets he had found evidence of at the Fraunhofer. They could measure atmospheric disturbances, but he had no way of replicating those experiments here.

He and Walter needed to be seen to be doing some testing. They were crossing Lake Seliger on a weekend beach outing one day, when Cocky noticed a line of buoys bobbing at the entrance to the small harbour. They asked the ferry skipper for one of the old buoys, and he gave them a split one. It was made of rubber, with a flat base, rounded sides and a pointed nose. Inside, it was a supposedly watertight cavity unless, like this one, it had split. They mended it with a patch, like a bicycle tyre.

Cocky and Walter built a water tank with a rudimentary wave machine in the workshop. The wave maker was a piece of wood that shuttled to and fro. They simulated turbulence by reaching right into the tank to sink the buoy, capsize it, and push it out of equilibrium. The casing for the gyroscopes had to fit, friction free, inside the buoy.

Helmut was not convinced that their experimenting with the buoy was time well spent. He was on the lookout for red-herring whims. His Russian masters had warned him to watch out for time-wasters. Seeing Cocky and Walter frolicking with a water tank made them appear prime candidates. Helmut would walk past, dodging splashes, until he demanded that they get on with some work.

Cocky asked to do his experimentation after hours, but this was not allowed. A new rule forbade after-hours working, and security around papers and materials was tightened even further.

Once, Cocky returned to the Gröttrups' home later than usual. From outside the house, he noticed a light in his own top-floor rooms. He ran up the stairs and found Helmut in his anteroom, busy examining his homemade radio set, which was a sprawl of wires and circuit boards. There was no casing, dials or switches.

'Good evening, Herr Gröttrup,' he said evenly. 'That's my radio set. Would you like a demonstration?'

'How do you tune it? There's no dial,' said Helmut gruffly, without any apology for being in Cocky's room without permission.

'You move this wire slowly that way. With this matchstick – don't use your finger. As you move it, you try to listen for anything that isn't interference or Russians droning on about writers or composers.'

'I'm going to have to confiscate it. It's against the rules to have communication devices. You know that. I could lose my position over this.' Helmut started to unplug it from the wall socket.

Cocky was desperate. The radio was his lifeline, his sanity, his hope for escape, his link with Eric, and his only entertainment. 'I've had it since I got here. It's only a receiver and not a transmitter – look – there's no transmitter here. I only use it for getting my music.'

Helmut raised a cynical eyebrow.

'I asked Irmgard about having it, for listening to classical music, and she said it would be fine,' said Cocky. 'I suppose I could come downstairs in the evenings and join the family.'

Helmut looked torn.

'Show me how you listen to music, without us hearing it,' he demanded.

Cocky plugged the set back into the wall. He switched it on to receive-only mode, and moved the matchstick so the wire moved slowly around. It took him a long time to locate a clear signal. He wanted to show that the device was barely capable of picking up anything at all, so he tried for a bit longer to convince Helmut he was trying to find some music. He skipped over Eric's frequency, and eventually, indistinctly at first, they could hear the strains of the London Philharmonic coming from the UK, playing some Vaughan Williams. The sound was tiny, so he plugged in the headphones to a receptor on a circuit board, and indicated to Helmut to put them on.

Helmut assessed the greasy and worn headphones with disgust, vaguely recognising the matted red fur, but he was unable to equate them with his wife's winter earmuffs. Cocky's new leather headphones were in the laundry bag in his wardrobe. Helmut placed the earmuffs gingerly over his ears. As he listened to *The Lark Ascending*, his expression slowly changed from discomfort to a surprised expression of delight. '*Toll*,' he muttered under his breath. '*Wunderbar.*'

Cocky waited anxiously for the piece to finish. Helmut stood with the headphones in his hand as the tinny, small radio noise continued. Cocky switched it off.

Helmut struggled to find the tone of voice he needed to ask a favour of another person. 'Can you build me a really small radio like this? With headphones?'

'Yes – we probably have the components in the workshop,' Cocky risked saying. 'Would you like it in a proper case?'

'I would like it in a proper case. Not just a circuit board, soldering and wires. I'd like an on–off switch. A frequency finder, and a volume control dial.'

'Ok, that's all possible,' said Cocky.

Helmut went whistling down the stairs.

Cocky slumped on the bed, heart pounding. He had been so close to losing his radio, losing his means of communication, and losing the trust he had built up so carefully with Helmut.

The next evening, Irmgard was busy in the kitchen, her arms covered in flour. She had bought a small amount of beef from the weekly market in Ostashkov. The market was a ferry ride away or a walk across the ice in winter. German housewives were allowed a regular shopping trip, under guard.

Irmgard was in the process of converting a rare piece of beef into a pie, supplemented with onions and a handful of mushrooms she and the children had picked. She often wanted to include Cocky in family meals, rather than having to take food upstairs separately. This was another area where she and Helmut didn't usually agree.

'Cocky,' she said, 'join us for our evening meal tonight. Look at this piece of beef. It's small but will chop up in a pie.'

'Thank you, I would love to, if that's alright with Helmut.'

'It's alright with Helmut,' said Helmut jovially as he came through the door. 'I have news about the Finnish quartet, Strava. They have agreed to play at our pre-Christmas concert! It will take place after the inspection visit by the Moscow delegation. A successful inspection followed by a brilliant concert.'

'I'll make those posters,' said Cocky.

'I'll get the women together to clean the clubhouse,' said Irmgard. 'And arrange for the quartet to stay over locally. It's just one night, is it, Helmut?'

'Yes, just one night, then straight back to Helsinki. They've got a busy season. With any luck, Irmgard, this will be our last Christmas on Gorodomlya – the next one could be in Moscow!'

The next morning at the workshop, Cocky was acutely aware that Walter hadn't shown up for work. He was waiting for Walter's help to prepare the demonstration of principles of inertial guidance. Cocky had explained to Helmut that the water tank was an approximation of a low-gravity environment, and he had reluctantly agreed to include the crude tank in the demonstration.

Walter was the main coder for the computer. He and Cocky worked well together, their expertise overlapping and finding in the other a ready sounding board for the conundrums they encountered.

'Where's Walter?' Cocky asked Helmut.

'Ah, Walter is ... not here.'

'But where is he? I need to know. It's important for the demonstration.'

'The Russians detected a radio signal from Gorodomlya – to a listening station in the United States,' said Helmut. 'Last night, they traced the signal to Walter's house, and they've taken him away to ... question him.'

'Question him?' echoed Cocky. 'What does that mean? Surely, they couldn't pinpoint a particular house. All the houses are set really close together.'

'Walter was found[1] with a radio set,' said Helmut. 'It looks like he had been using it to pass details of our top-secret work direct to the Americans. He was always asking me about our developments, always interested in every detail. I feel responsible for the fact he knew so much. Then it looks like he he was transmitting to the scientists at the Americans' rocket project.'

'Walter didn't ... we don't know anything, though,' said Cocky. 'The Russians don't tell us everything.'

'They'll be even more guarded now,' said Helmut. 'And it might delay their decision about when we can leave the island. It's already gone far beyond their promise of release after five years. They won't want any more of us taking our knowledge straight to the Americans.'

'What will happen to Walter?' asked Cocky. 'Will he "disappear"?'

'Not until they have every little bit of information from him that he passed on. Then he will be punished as a spy.'

'Punished as a spy,' breathed Cocky.

'He will probably be sent to a labour camp for a few years, maybe in Siberia, or ...'

Cocky blanched at the thought of kind, funny and sensitive Walter working in a labour camp, or worse. Labour camps were known to work people to death. 'I need him to finish programming the analogue computer for the demonstration. Without Walter, we won't be able to show the Russians the progress with the guidance system.'

Helmut backpedalled, shocked at the possibility there would be nothing to show the Russians. 'He's only in Ostashkov ... at the moment.'

'Get him back,' said Cocky, 'or the whole demonstration will fail.'

'I'll go and find him ... tell them there's been a mistake, and bring him back.'

'Hurry. We've not got a moment to waste.'

When Helmut had gone, Cocky undid a section of the coding that Walter had completed. He spent the rest of the morning in a state of anxiety over Walter. He resolved to stop making transmissions to Eric in Germany. If the Russians were detecting radio signals from the island, those from Walter would now stop, and whoever was detecting them would turn their attention to the signals coming from his own radio.

Eric had passed a minimum of information to him from the Americans about their inertial guidance system; details that Eric gauged would help Cocky make just enough progress. For his part, Cocky mentioned to Eric as much as he could glean from discussions and observations made during inspection visits and in the day-to-day work.

By six o'clock, Walter and Helmut had still not materialised.

The next day, Cocky had to replace Walter's coding with some of his own, and he tested the computer's interaction with the water tank. By the end of the week, he could generate results that showed the potential for inertial flight control through a turbulent environment.

There was no satisfaction in any of it. He grieved for Walter, and could find nothing to say to Helmut, who had come back alone from Ostashkov, saying that Walter had already gone.

The Inspection

The day of the visit from a delegation of Russian scientists from the main Moscow institute was bright and cold. Cocky had his demonstration ready to show. It was a muted success, with additional improvisations by one of the Russian visitors, who plunged his arms enthusiastically into the water tank to agitate the buoy, getting his sleeves wet in the process.

The prototype inertial, or gyroscopic, solution was cautiously accepted as a potential method for the guidance system. It was an entirely different way of guiding a missile from the high frequency radio wave method favoured until now by the laboratory. Cocky was uncomfortable with the thought that due to Eric's drip-feed of knowhow, he was in danger of raising his own profile.

Cocky was instructed to work with the Russian, Viktor Ivanovich Kuznetsov.[2] Helmut was relieved that the Russians had the developmental step they wanted from their visit, and one of the delegation had remarked that the demonstration was 'an early Christmas present' to the project.

Cocky's mood remained low. He couldn't enjoy the moment if Walter wasn't there. He feared for Walter, and there could be no relief from the sick feeling in the pit of his stomach about his friend's fate.

The Big Freeze

The pre-Christmas concert at the clubhouse came as a welcome distraction from Cocky's searing anxiety. The Finnish quartet had arrived, and the talk was that although they had rooms in a hotel in Ostashkov, they had accepted an invitation from a family on the island to share their evening meal. People tried to catch a glimpse of them, especially the children, and there were reports that one of the musicians, a young woman, had yellow-blonde hair, 'like a fairy princess'.

The applause was loud and long when Strava walked out onto the small stage, dressed in concert black and holding their instruments. The Russian delegation had the best seats, and the children had the front rows. The lead player held a beautiful violin; next to him was another violin, then the viola player and the female cello player. She radiated serenity. Even the women in the audience were fixed on her. The three men had longish hair and beards, and lively, innocent faces – so different from the Russian look.

The concert began and a silent audience watched and listened. The music was urgent, original, full of discordance and surprise, bursting with fearlessness. The quartet communicated with each other with small glances and nods, keeping in time and throwing themselves into the piece with abandon. It was mesmerising, and Cocky almost forgot about Walter. The music sounded infinitely richer in reality than through the radio.

At the end of the performance, when Helmut had stood at the front to thank the musicians and the applause had finally faded, Cocky stayed in his seat, not wanting to return to reality. The young cellist came straight up to him from the stage and held out her hand towards him.

'Kaarina,' she said confidently with a wide smile. 'And you are?'

'Neville,' said Cocky, instantly dropping his nickname.

'Neville ... what is that? An English name?'

'Yes, I'm English, though most people here are German, or Russian.'

'Ah, the Russians,' she mused. 'It is they who asked us to play.'

'Yes, but I found you on the radio. An interview with Eino Rautavaara,' said Cocky.

'We like to play Eino's music,' said Kaarina. 'He is talented. Very special.'

'I love his work,' said Neville.

'You like classical music?' she asked.

'Yes, very much. I was always going to concerts in England, but haven't been to anything since the war.'

'You must come to Finland,' she replied, with sudden warmth. 'We will go to concerts, have some nice times.'

'I can't,' said Cocky. 'I'm a prisoner here. On this island. I – and all of the Germans, in fact – are unable to leave. We are scientists; we are forced to work on the Russians' next rocket.'

Helmut was approaching them, so the conversation had to stop there, but a shadow passed over Kaarina's lovely face.

'We've got a problem,' Helmut was saying, as crowds of excited people surged back in through the same door through which they had just left. 'The lake has iced over, so the steamboat back to Ostashkov can't leave.'

'We have to leave,' said Kaarina. 'We have a major concert in Helsinki at the end of the week.'

'*We* have to leave,' said the Russians. 'We have to report to the Kremlin tomorrow morning.'

'We will have to sleep here at the club,' shrieked some local children, running off to find a classroom cupboard, where there were mats and blankets.

The Russians immediately radioed for a helicopter. A bulbous seaplane arrived an hour later, landing on the snow-covered clubhouse lawn. The Russian delegation filed aboard it, and it rose into the air and disappeared over the horizon.

'Can you organise another one for the quartet?' Helmut shouted to their departing backs, as they jostled to get on board. But the Russians had not answered. Helmut realised that his influence didn't extend to ordering a military vehicle to take the quartet back to Finland.

Kaarina and her three colleagues clustered around their packed-up instruments, with their warm coats and colourful hats on. No one dared to tell them that unless a helicopter or seaplane could be found, they might be unable to leave until the ice thawed.

The steam ferry skipper pushed his way into the room. He had been trying to break the ice around the ferryboat with a pickaxe, but it was freezing solid again as soon as he managed to make any impression. 'The lake has frozen over,' he declared. 'The ferry can't move. It's not safe to try to walk on it. Remember what happened to Olga Brodanovic and her dog? We will wait until morning, and then see what to do.'

The quartet stayed put, stubbornly holding their instruments. They had agreed to come to this island for one evening, with an overnight stay in a hotel in Ostashkov. A chauffeured car would take them back to Helsinki the next morning.

Cocky went to speak to Irmgard, who had overwintered on the island several times before. Everyone knew the lake would freeze over at some point, usually in late October or early November, taking about two weeks to fully ice over, but this year the freeze had come late and swiftly. He felt responsible for the musicians getting home, aware that he and Irmgard had pulled the strings for their visit.

'The car organised for the quartet – will they be able to meet with it in the morning?'

'I don't know,' she replied. She thought the roads would be just as iced over as the lake was; even if they wanted to brave walking over it to meet the car, it would be too great a risk to their instruments.

'Where can we put them for the night?' asked Cocky. 'They can't get back to the hotel. They'll have to sleep here. Not with the children. Somewhere a bit private?'

'The boiler room is the driest and warmest place,' said Irmgard. 'We could get some blankets and settle them down there.'

The quartet were becoming restless. They had expected to get home to Finland straight afterwards for their own musical run-up to Christmas. They didn't want to sleep in the boiler room and were demanding a different solution. Irmgard stood with them and explained again that in her experience there was nothing they could do right now and a solution would present itself in the morning.

Cocky selected the cleanest blankets and cushions from the pile and went up and down the steps, creating four sleeping places in the dark and noisy boiler room. Eventually, the quartet members carried their precious instruments down to the boiler room and sat gingerly on the blankets. Cocky put on the kettle and Irmgard found a pack of cards in a dusty cupboard. He rummaged in the same cupboard and found a half bottle of whisky wedged securely at the back, where someone, perhaps a clubhouse schoolteacher, appeared to be keeping a hidden stash. People were wandering into the kitchen to make coffee, while Cocky, Irmgard and the four musicians drank whisky and played cards, trying to make the best of a difficult situation.

'This is schnapps,' said Kaarina, 'not vodka.'

'It must have come over with us Germans, a few years ago, because you can't buy it here,' said Irmgard.

'It's good schnapps,' said Bjorn, drinking it back in one go.

'How come you are all living here, on this strange island, surrounded by barbed wire and guards ... with children and families here?' asked Kaarina.

Irmgard explained how the Nordhausen Institute in Bleicherode had been relocated to Russia in 1946. (Cocky hadn't heard the event referred to so positively before.) Irmgard was proud, saying, 'My husband Helmut was Director there.'

'I think Korolev was there too?' queried Cocky.

'Yes, until one day in October 1946. The Kremlin brought the scientists they'd gathered in Bleicherode onto Soviet soil. With their families. They loaded everyone onto trains. In the early morning, when it was still dark. After a heavy party, when people were struggling with hangovers. Except me – I wasn't drinking that night because of the children.'

'Did the Russians get people drunk on purpose?' said Kaarina.

'Helmut was hardly able to stand,' said Irmgard. 'I had to do all the packing up.'

'What a cheek!' said Bjorn. 'Taking all those people and their families, to … to make them … to force them to do … what, exactly?'

Irmgard explained that Helmut had written a protest letter when he was actually on the train, but the Russian response was to reassure them it was only for five years, unless they decided to stay.

'What sort of scientific work?' asked Bjorn gruffly, who hadn't taken off his woolly hat, despite the warmth of the boiler room.

'I can't, obviously, tell you what sort of work,' said Irmgard. 'And we can't escape. They have our passports, and the perimeter fence goes around the whole settlement, including the forest. The town on the far shore, Ostashkov, is where we wives go to market, but under guard. We are counted out and counted back in.'

'Irmgard?' came Helmut's voice shouting down the stairs, and Irmgard had to go, leaving her whisky untouched, which Bjorn snapped up immediately and downed.

'Has anyone ever tried to escape?' asked Miko, when Irmgard had gone.

Cocky thought about Walter. 'I don't know,' he said. 'Anyone who tries to escape, or works against the Russians in any way, can easily be "disappeared". No one knows where these people go. Maybe to labour camps, or made to work in mines or build roads and railways. Or, they can be casually murdered.'

'What about you?' Pekka asked Cocky. 'You're not a German scientist.'

'I'm a British electronics engineer.'

'What are the Russians doing all this for, though?' asked Miko, who had been lying back on his makeshift bed, looking like he might be sleeping.

'Since Hiroshima, the Soviets have feared the Americans' powerful weapons, their atomic bombs,' replied Cocky. 'They are hell-bent on developing their own, keeping up with the Americans. In their view they are balancing out military power across the world.'

'What about rockets?' Kaarina asked.

'The rockets are the carriers, to get a bomb to a target without using human pilots,' said Cocky. 'The Soviets copy rocket technology they admire. They are making this their national priority.'

'A rocket – with an atomic bomb attached to it,' said Kaarina.

'Oh, God,' said Miko.

'The lady – Irmgard – wouldn't tell us what the top-secret developments are,' said Pekka. 'But you just did.'

'You just said you've invented something the Russians need for their rocket,' said Bjorn. 'Why would you do that? If the Russians want to attack other countries?'

'Finland, Britain, America …' said Miko.

'I am working on something their rocket needs,' explained Cocky. 'I was hoping never to pass it over to them. Today, though, I had to give it over to the Russians. But I was forced to come here. I had to work on it.'

'That's appalling,' said Kaarina. 'You, Neville, are helping to create something that will kill people. *Kill* people. With terrible bombs.'

'What do you mean, they brought you here?' demanded Bjorn.

'I was abducted, from an airfield in Germany,' said Cocky. 'Pushed into a SOXMIS car when I got off a military plane. Put on a flight to Leningrad, then by car to here.'

Pekka and Miko looked sceptical.

'My colleague, Walter, was "disappeared" just last week, and that will happen to me. I'm using a radio as well. I need, more than anything, to get away from here. Leave the island. Leave the work. I want no part in it, and I never have wanted it.'

'Why were you *abducted*?' asked Miko, cynically.

Cocky put his playing cards down, as the others had done, and took a sip of whisky. 'In February 1946, I had just joined the RAF when I ended up on a scientific mission to Austria, to an institute called the Fraunhofer, where the Germans had been experimenting with radio control systems for the V-2 rocket. I met Korolev there, the man who is now running the Soviet rocket-building project. But at that time, he hardly seemed to know anything. He pinned me in the kitchen once, trying to find out if I could be useful. But he didn't have his heart in it. After that, I was posted to Singapore.'

'You didn't get deported with all the Germans?' queried Miko.

'No, they were moved out here, and other places around Moscow, later that year, October 1946. I wasn't targeted until more than four years later. After

Singapore, I was working at a Signals Unit in Germany. I'd flown back to the UK for some equipment, and as I just told you, when the plane touched back down on the airfield in Germany, the NKVD[3] were waiting for me.'

'My God,' said Kaarina.

'Why you, though?' said Pekka.

'It's taken me ages to work that out,' said Cocky. 'The only time I came into contact with them was at the Fraunhofer in 1946. They know I'm an electrical engineer. That I worked at an aeroplane factory. That I do technical drawing. You know Helmut Gröttrup? Irmgard's husband – the lady who was here just now? I lodge at their house. I work in Helmut's laboratory here. Helmut thinks I'm a radio guidance specialist.'

'Are you?' asked Pekka.

'It's a big disappointment to him that I'm really not,' said Cocky.

In the warm and gloomy boiler room, the musicians shifted positions to make themselves more comfortable. Cocky was desperate that the group, particularly Kaarina, should not think he had chosen this work.

'I think that Helmut, or his workshop manager, asked the Moscow team for a guidance specialist to join the work here on Gorodomlya. Helmut told me he wanted a particular one, Viktor Kuznetsov, to transfer from another workshop, but it wasn't allowed. Someone at Podlipki may have put me forward – I don't know.'

'But you're not developing radio-controlled guidance?' asked Miko.

'No, I'm working on inertial guidance,' said Cocky. 'It's more reliable. I don't want to give away knowledge about it. But today, I did have to. Before the concert tonight, I had to present my work, and Kuznetsov was there.'

'I don't think Helmut will get rid of you,' said Miko. 'Not if you've invented something they need.'

'I've not invented it. I'm just applying a different idea to an old problem. The main Russian effort is still going into radio control systems. Even if the inertial system suddenly becomes highest priority, it's being taken off me. As I mentioned, the person I was working with, Walter, was "disappeared" last week.'

'Who's Walter?'

At Kaarina's simple question, Cocky's voice broke as he talked about his friend and colleague – kind, funny, uncoordinated Walter, so unsuited to hard labour, so undeserving of punishment.

'Walter would have loved your concert,' he blurted. 'He was …'

Kaarina reached for his hand, so Cocky used his other hand to wipe his face.

'Eino believes that music can build bridges between nations,' said Kaarina. 'He was excited about us coming here to do the concert, at the invitation of the Russians.'

'Music is common ground,' agreed Cocky.

'We will help you,' she said gently. 'We will help you escape from Gorodomlya.'

Bjorn, Pekka and Miko looked concerned about the promises that Kaarina was making.

'I only want to get myself out of this godforsaken ice hole,' said Miko.

But Bjorn and Pekka sided with Kaarina.

Cocky woke to the sounds of schoolchildren searching through the kitchen cupboards at the top of the boiler room steps. His head was pounding and his fingers still intertwined with Kaarina's.

Escape Plans

Cocky jerked his hand away in confusion. The quartet's three men were sprawled on piles of bedding in the half-dark of the boiler room. They appeared to be asleep, but anyone could have noticed his hand, resting on Kaarina's through the night. He got up quietly, making his way up the concrete steps and through the throng of children.

A small knot of people had gathered outside the clubhouse. Irmgard had used the club's telephone to contact local Russians she knew from the guarded shopping trips to Ostashkov, beyond the barbed wire fence. Now, their emergency response vehicle was arriving. In the distance, over the sparkling new ice, an old-fashioned dog sled could be seen coming across the lake. A newly frozen body of water was an unknown entity. Patches of thin ice meant a person or vehicle could plunge through – local people were all too familiar with such hazards.

The group clutched ropes as they watched the sled approach. It was ancient, wooden, with peeling red paint. The four old dogs had matted coats, and one of them was limping. The Russian sled driver, with his two remaining teeth and sporting a fur hat with earflaps, was similarly ancient. Irmgard greeted him in Russian, kissing him on both cheeks and making sounds of gratitude. He gestured to her about the thickness of the ice over which he had just travelled. She was reassured, and her two children were the first to sit in the fur-lined bucket seat. The driver made a wide arc with the sled, before depositing them safely back on shore. After that, priority was given to the Russian children and adults from Ostashkov, with never more than two people aboard.

The quartet members still expected an official car to meet them at Ostashkov, but there was no way of knowing whether this would materialise. Kaarina had to go alone with her cello, as it was the size of a child. Cocky thought how beautiful she looked as she set off across the ice with her instrument. On the next trip, Cocky, Bjorn and his violin shared the bucket seat. Irmgard and Helmut were left at the clubhouse.

Bjorn wasted no time in taking the opportunity to have a private conversation with Cocky. He had been waiting for this moment, aware that Cocky had joined the ranks of Kaarina's admirers.

'Kaarina is the girlfriend of Eino – our friend, Einojuhani Rautavaara, the composer. It's serious. For months, she waited for him to return from New York. Her heart is not free.'

'I think she feels sorry for me,' said Cocky miserably. 'Being here. Being made to stay here.'

'Yes, that's right, she feels sorry for you. She is a very kind person. We are all, all of us, a little bit in love with her. She is our sister, our mother, our friend, the bright star in our ensemble, our glue. She keeps us all together.'

'I can see she is important to all of you. I'm sure that Eino must be waiting anxiously for her return. Kaarina only wants to help me escape, that's all.'

Bjorn was satisfied he had blocked any further moves from Cocky in Kaarina's direction. Their hands touching through the night in the boiler room could have been accidental, just the way they happened to fall asleep.

'We will all help you,' said Bjorn. 'When we get back to Helsinki, we will ask our government to put pressure on Stalin to release the scientists of Gorodomlya.'

134 The Gorodomlya Island Project

Cocky considered this. It came from Bjorn's kindness, but being released by Stalin wasn't a realistic prospect. 'There are many, many prisoners,' he began, 'thousands of Russian people, anyone who has ever voiced dissent about Stalin's rule, or is simply clever enough to do so.'

Bjorn felt that Cocky hadn't understood his own country's experience. 'The Soviets invaded Finland in November 1939. The Winter War. Three months later, we had defeated them, and we signed the Moscow Peace Treaty. They tried to make us part of the Soviet Bloc, tried to enforce communist rule on us. A dark, dark time, but we fought back and won, and Finland is now a proud, independent country.'

Cocky saw that Bjorn and the other members of the quartet had come to Russia, against their political instincts, to forge cultural links based on music. But the Russians were not responding in kind, and hadn't even extended the courtesy of getting them safely home.

'Stalin', said Cocky quietly to Bjorn, sure that no one could overhear, 'holds thousands of men, in closed settlements all over Russia. I heard about a whole city being built to manufacture plutonium, using Polish prisoners of war. He's hardly going to release me, or the Germans.'

Bjorn embraced his violin as he settled into the sled, gaining comfort from its benign, civilised presence. More than anything, he wanted to get back to his own, free country. He understood that Cocky had asked him for his help to avoid becoming the next person 'disappeared', but he felt powerless in the face of the Soviet regime.

The opposite shore came into view. 'We will ask for the diplomatic release of just you,' said Bjorn. 'One British person.'

'The British prime minister, Winston Churchill, knows that I am here,' said Cocky. 'He might remember me. It's worth asking him. If you can do that for me, Bjorn.'

Bjorn looked askance at Cocky's arrogant-sounding hope that Churchill would remember him. The idea of asking the UK government for the diplomatic release of his friend was taking shape in his mind. He would approach the British Embassy in Helsinki, but he doubted that it would be easy.

Bjorn and Cocky reached the icy shoreline and clambered stiffly out of the sled. Bjorn shouted his thanks to the driver before making his way to where his colleagues were waiting. There was no sign of the car for their onward journey. Cocky needed to report to Irmgard that the car had not yet arrived, so he set off back to the island in the sled, which was, by now, one dog down.

Cocky spent the next hour at the Gröttrup house, helping the Russian domestic worker to make a stew for when Irmgard and Helmut got back from the clubhouse. It was almost fun, combining sparse ingredients so the small amount of meat would stretch. The children too were genuinely helpful, and Cocky was pleased with his first ever attempt at cooking.

The stew was very welcome to the family. There were anecdotes about the night at the clubhouse and the fun of the dog sled. Only Helmut was taciturn. The classical music evening had not brought about the cultural connection with the Russians he so keenly desired. It was embarrassing that he couldn't command a helicopter to take the Finnish quartet home. To top it all, the young cellist had taken no notice of him at all, preferring the company of his not entirely welcome lodger.

Towards the end of the meal, Kaarina, Bjorn, Pekka and Miko appeared outside the door, looking dishevelled, and knocked on the window. Their car hadn't arrived, they couldn't get to the Ostashkov hotel, and the sled driver had had to bring them back over the ice to the island. Irmgard invited them in for stew, and afterwards outlined to them their options, as she knew from experience that the ice was unlikely to melt any time soon.

'The train?' queried Kaarina. 'Someone said German workers were brought here by train.'

'Those were specially commissioned trains,' said Irmgard. 'There's a night train, but that only shuttles between Moscow and Leningrad; it doesn't stop here.'

Witnessing the frustration of the Finnish quartet was difficult. Cocky racked his brains for a solution. The next day, he set off to visit them in a house Irmgard had organised for them by asking two German families to move in together. Bjorn met him at the door and invited him in, offering a tot of Finlandia vodka. Cocky declined the alcohol, asking for tea instead. He wanted to keep a clear head.

'Helmut was here just now,' said Bjorn. 'The local railway, which usually keeps going throughout most of the winter, has been out of action for nearly a year. They are electrifying the line. So that possibility is out. The roads are iced up and impassable by car, for at least a few weeks – maybe a couple of months. We need to wait for the thaw, and then for the worst of the mud to clear. Helmut says it comes down from the hills like rivers of sludge.'

'Is he going to ask again for a military helicopter?' asked Cocky.

'He's not sure about that,' said Kaarina. 'I don't think he's got the authority.'

Pekka kicked the door. 'This is a nightmare. He promised to ask again for a helicopter.'

'Helmut seems to think that our only course of action is to wait the few weeks,' said Pekka bitterly. 'But we don't want to be stranded here.'

'We've just had to send a telegram to Eino, cancelling our next concert,' said Kaarina sadly. 'This is doing us no good at all professionally.'

'Being stuck here is no good for anyone's career,' said Cocky.

'I'm sorry. I'm complaining about being stuck here for a few weeks, but you've been here for – how long?' queried Kaarina.

'Nearly fifteen months,' said Cocky. 'But the Germans have been here for six years, since the end of 1946. I was only brought here at the end of 1951.'

'Why are you so concerned about being "turned over to the Soviets" like your friend Walter', asked Pekka, 'if you were brought here for a specific purpose?'

Cocky took a sip of the Finnish vodka, as the tea hadn't appeared. The vodka was much smoother than its Russian equivalent. 'I have to do the job they brought me here to do. But I needed help with aspects of the design, from my boss in the RAF, Eric. He's based in Germany but he's in touch with the Americans. There's an enormous effort going on in the USA – it's like a race to build the most accurate rocket.'

The quartet was quiet.

'So I made a radio,' he said.

'You made a radio,' repeated Pekka. 'So what?'

'I'm getting design information through on my homemade radio,' said Cocky, 'from a colleague.'

'That would upset the Americans, not the Russians,' cut in Kaarina. 'In fact, if that's the case, you're going above and beyond to help the bloody Russians!'

Cocky was appalled. He needed Kaarina to understand, to be on board, or she and the others would never help him to escape.

'Wait,' he said. 'There's a fine balance here. I need my colleague's help to make progress, to stay key to the project and not be "disappeared" like Walter. Eric passes on to me pointers on how certain design challenges are being tackled elsewhere. But, we work out how the process can be slowed right down, how key facts can be hidden, how a set of results can appear correct, but are in fact very slightly, imperceptibly, wrong, so that the system won't work in practice. Believe me, Kaarina, everyone, we don't want Stalin to have a powerful, controllable rocket in his hands!'

Kaarina breathed out, and the tension of the room dissipated significantly. People took another sip of vodka.

'And all the time we're doing this, I am passing on to my colleague intelligence that could be valuable to the Americans. Using my radio. I'm picking up details of how the Russians are getting on – mostly just small details, occasionally something like a new Kremlin decree. It's hard because they're determined *not* to share anything with us. Every time there's a visit from the main laboratory near Moscow, I'm around, I'm in the background, listening – picking up on the personalities, the fallouts, the areas of work that are stuck, which components can't be sourced. It's day-to-day stuff, but it could be of some value.'

'So, there's a trade-off. In return for some technical help,' said Pekka.

'What is your project?' asked Miko.

'An onboard, inertial guidance system,' said Cocky.

'Oh,' said Miko.

'It's not brain surgery,' said Cocky.

'It is rocket science,'[4] laughed Kaarina, and the remaining tension was broken. Their laughter reminded them that actually, they were friends.

Cocky needed to make just one more point. 'What I'm doing, with my radio, is exactly the same as Walter was doing, passing intelligence to the Americans, using the same method – a homemade radio. As soon as Helmut works it out …'

'Does Helmut know about the existence of your radio?' asked Kaarina.

'Yes, I showed him it. He listened to some classical music, some Vaughan Williams.'

'You showed him it? I don't understand why you would do that. That's stupid,' shouted Miko.

'Helmut found my radio, in my room. I convinced him it was only a receiver, and that I used it to listen to classical music. It is usually just a receiver. It's really small. I found you on it – I found Einojuhani and Strava on it – and it's true, I use it to listen to my favourite classical music. That radio is my lifeline to normality.'

'When you switch your radio to transmitter mode,' said Pekka, who knew a bit about radios, 'you get hold of this – Eric – and pass on intelligence. How often?'

'I've stopped now but it was every couple of weeks. Eric can choose to take it further, to the British and American governments.'

'I suppose the Soviets consider that spying?' Miko suggested.

'From their perspective, that's exactly what it is,' said Cocky.

'You don't know what happened to Walter,' soothed Kaarina. 'He might be alive somewhere in Russia, building a road or a railway.'

'I hope so,' said Cocky. 'I really do. People don't usually *reappear*, though. Their relatives have to forget them. People aren't allowed to mention the names of "the disappeared" in public.'

As they sipped their tea, there was a knock at the door. It was Renate, bringing cabbages and bread. Renate's baby was tucked into her coat, his small red nose and huge eyes visible beneath his fur hat.

'Come in,' said Kaarina, and the two women chatted for a while over the baby's head about the prospects of the quartet getting home.

'Dog sled back to Finland?' quipped Miko.

'We can't stay in this house – the family who live here thought we'd only need it for one night,' said Kaarina.

'If you're still here next Saturday, some of us are going skating on the little inland lake. My husband makes skates out of old boots – you can borrow mine. I'm not risking it this year because of the baby.'

'We'll be home by next Saturday,' said Miko. 'Thanks anyway.'

'Thanks, Renate,' said Kaarina. 'We just might have to join the skating party.'

Renate left in a flurry of warm goodbyes and waves at the baby.

'Can you get your passport somehow?' Kaarina asked Cocky. 'And just leave when we do?'

'Passports are in the guard's office by the gate,' said Cocky. 'Mine's a military one.'

'We just have these visas,' said Bjorn, fishing his out of his pocket. It was a piece of pale pink folded cardboard, with a photo and some printing. 'Not hard to copy. I've got a camera. I can make a Finnish visa for Cocky.'

'A Finnish name, then, is needed,' mused Kaarina.

'Arvo or Paavo,' said Bjorn.

'Join our quartet! Make it into a quintet!' laughed Kaarina.

'I can't play anything. Only the comb,' said Cocky.

'The comb?' queried Miko.

Cocky pulled his comb out of his pocket, and put a piece of greaseproof paper over it from the cake wrapping. He tried humming into it, and the sound amplified into a tuneless buzz.

Pekka began singing a Finnish folk song, to divert from the sound of the comb. Miko and Bjorn started to drum on the table and sing along too – a sad song about rivers, meadows, and never again seeing one's true love.

Kaarina emerged from the back bedroom with her cello and drew out some long, mournful notes. She was shyer playing here than she had been on the stage at the clubhouse. After a few minutes, there was a tentative knock at the door. A group of children came in, banging the snow from their boots politely – the same ones who were always trying to catch a glimpse of 'princess' Kaarina. Another folk song started, and Bjorn reached for his violin, adding in a soaring, sweet sound that blended with the cello. Cocky answered the door to a group of the children's parents. He pushed the furniture back to make room, and found glasses and mugs for drinks they had brought. The song was now lively, reminiscent of spring weddings. A few children were dancing.

Renate came back with her husband and baby, and Helmut's family arrived. Irmgard twirled her daughter around in the small space. A German husband asked for a folk song from Germany, but ended up singing it himself while the musicians picked up on the tune. Some songs were so similar in both cultures that each nationality sang their own version, laughing about the coincidences of melody and rhythm.

Bogdana and her husband came in, pushing into the room through the crowd. Cocky recognised Bogdana's husband as the dog sled driver. He was offered drinks and toasts in honour of his timely help with transporting people over the ice.

The family who lived in the house arrived, and joined in tentatively, worrying about breakages. Their two daughters ran around the house finding where Kaarina was sleeping, and the elder one was excited to find that her room was the one chosen. 'Two men are in yours,' she told her younger sister.

'If we could have one room back,' said the parents, 'we could put the girls on the settee in here, and move back in.'

'I've got a sitting room free,' said Cocky. 'In Irmgard's house. For one of the men.'

'That's fine,' said Irmgard, 'if you don't mind sharing, Cocky.'

'I'll come to yours,' said Bjorn. Irmgard looked delighted; she loved being hospitable, and she liked Bjorn.

'It's not for long,' said Pekka. 'Just until the ice thaws and the roads clear.'

'Bogdana told me that the freeze can last for *three months*,' said Kaarina.

Cocky and Bjorn set off to the attic rooms of the Gröttrups' house. In the little sitting room, Bjorn took an experimental photograph of Cocky's face against the white of the wall, and several more of his own visa. The ageing horsehair settee was too short to accommodate Bjorn's long legs, so Cocky gave him the bed while he took the settee. It had been a wonderful evening, the best since arriving on the island.

In the morning, Cocky and Helmut set off to work, walking gingerly along icy tracks, while Bjorn experimented further with the visa before meeting up with the other quartet members to explore the snowy forest on cross-country skis.

'What do you make of this place?' said Miko. 'It was once land belonging to the monastery, but there are no traces of monks here.'

'Stalin forced all the monks out, in 1928,' said Miko. 'And there's a local legend I heard last night, that one of the monks drowned himself in the inland lake. His ghost still wanders around, chanting.'

No one responded to this, but the pathos of the island being turned from a peaceful religious sanctuary to a place that served the interests of war was not lost on them. They skied for a while in silence, as clumps of snow fell from the trees, and the low winter sun flashed between the trunks.

'I can't work out why the institute building is so large and well built,' said Kaarina. 'It's really extensive. I heard there's even a wind tunnel.'

'Ah, yes, I heard something about that last night, too,' said Miko. 'It used to be for biological research, supposedly for curing foot-and-mouth disease. You know – cattle die of it. Then, it was a chemical research laboratory that experimented with chemical weapons, and deadly viruses.'

'My God,' said Bjorn.

'There's been a lot of new apartment building near the institute. Did you notice it?' Bjorn asked the others. 'No one's sure why apartment blocks are going up, when the Germans are expecting to get repatriated within the next year or two.'

'Why doesn't Cocky just ask to be repatriated?' said Miko. 'Irmgard said that some groups of workers had already been allowed to return to Germany – to East Germany, that is.'

'It's not up to him,' said Bjorn. 'Don't you realise, the workers here are made to stay for as long as they're useful? It's only workers they consider replaceable that get to go home.'

'What is Cocky actually doing for them?' asked Miko. 'I didn't get it.'

'He called it an "inertial guidance system",' said Bjorn. 'From what I can gather, it's needed for their latest rocket, but now he has to transfer his knowledge to a Russian called Kuz-something, and once he's done that, he will be dispensable.'

'"Dispensable" could mean he gets sent home,' said Miko. 'Back to England.'

'It's more than that,' said Kaarina. 'He thinks that Helmut will find out, or already knows, that he's been using his radio to pass information to the Americans. Helmut could be waiting until he's passed on his knowledge to

Kuznetsov, then, just like Walter, he'll "disappear". In my opinion, he has to get rid of that radio.'

'You're staying in his rooms, Bjorn. Have you seen this radio?' asked Pekka.

'No. I'll ask him where it's hidden. I'm on with creating the Finnish visa for him, though.'

'Are we really going to do this?' asked Miko. 'Help him escape? Can we be sure he is who he says?'

'He's definitely British – no other nationality has that sense of humour … would play a comb, so badly, with professional musicians,' laughed Bjorn.

'He's awkward, uptight, repressed,' said Pekka. 'Very British.'

'He's stressed, not repressed,' said Kaarina. 'He is upset about his friend Walter.'

'I think he was dwelling on it a bit too long … because Kaarina was holding his hand,' teased Pekka.

'I believe him,' said Bjorn. 'His story stacks up. He told me how the Russians abducted him from an airfield near his RAF station in Germany. They took him up the Brocken, then flew him from Berlin to Leningrad. He couldn't have made all that up.'

'He certainly could have made it up. It would be best if we could check his ID,' said Miko.

'I believe him as well, because he's not very good at lying,' said Bjorn. 'He's not exactly cool and smooth.'

'His ID is locked in the guard's hut,' said Pekka. 'We could break in, but that would alert people. We might have to trust that he is who he says he is, that he's a scared Brit from the RAF.'

'We need to work out a whole new Finnish ID,' said Bjorn. That would get him as far as Helsinki. Out of Russia.'

'I think we really do need to become a quintet,' said Kaarina.

'He'll never learn to play an instrument in the time we've got,' said Pekka. 'He's not even musical.'

'He likes listening to classical music. And we only need to look like a quintet, not sound like one,' said Kaarina. 'We won't be performing.'

'He could buy an old German violin from one of the community orchestra musicians,' suggested Miko.

'He would need to grow his hair, even try for a beard,' said Pekka. 'And get a woolly reindeer hat.'

'He'll need a few words of Finnish,' said Kaarina.

'And this visa,' said Bjorn. 'He'll need that, and it will need to be perfect.'

'I've been thinking about the music titles. We could try wood-carving the 'i' and the 'n' and printing with them, matching the blackness of the ink,' said Bjorn. 'Change qu*a*rtet to qui*n*tet.'

'Not even the most conscientious border guard would check things out to that extent,' laughed Kaarina. 'How many border guards even know what a quintet is?'

'We can't take any chances. If there's something, any small thing we can do now to make our story believable under scrutiny, we've got the time to do it,' said Bjorn.

'We've got plenty of time,' said Kaarina. 'In fact, time is all we do have.'

A profound peace permeated the forest, brushed only by the swish of their skis moving through the trees. In the distance, across the frozen lake, the distinct bang of explosions rudely interrupted the moment. On the horizon, rocket shells could be seen, miniature ones, being fired into the air.

'They're not going up from Gorodomlya itself, they're coming from another island, further away,' said Miko.

'There are loads of islands on this lake. Maybe a hundred. The further you go, the more you see,' said Bjorn.

'Experimental rockets. Cocky told me about them,' said Kaarina.

'I think he likes you, Kaarina,' said Pekka.

'I like him,' she said.

'He really *likes* you, though,' pushed Pekka.

'I've got Eino. We're getting engaged. When I get back. I'm sure of it.'

'Eino is in love with his music,' said Pekka.

'He's in love with me too. I don't mind coming second to his composing. He's just brilliant.'

Saturday came, the day of the ice-skating party on Gorodomlya's internal lake. There weren't enough adapted boots to go round, so people were sharing. The adults formed twos and threes. Cocky had to take turns with Helmut and Bjorn, and quickly realised he had no natural grace. He was too fearful of falling to give it a proper chance. Irmgard and Kaarina were sharing Renate's skates, and each of them made a go of it in their own way. Irmgard held Renate's baby while she skated. At lunchtime, there was a picnic of bread, cheese and apples.

Not long afterwards, the watery sun started to set. Cocky had hardly left the edge of the ice during the party, and was relieved to be able to stop his floundering. Kaarina skated past him confidently, doing her last few circuits, and he tried once again to master the skates, copying her moves. He gathered his courage and set off boldly across the ice in her wake, eyes fixed on a strand of bright hair escaping from her hat. He felt like he was flying.

'Good, Neville,' she said to him when they stopped. 'You've got it.'

They each picked up a handle of the picnic basket and set off back towards the houses, with Cocky feeling exhilarated over his unexpected skating prowess.

'Tell me about your life in England,' said Kaarina.

'I'm one of nine children. Only six of us survived infancy. One, Bernard, died when he was 4. When we were children, all our names began with B.'

'What was your name?'

'Bruce.'

'You changed it?'

'Most of us switched to our middle names. Except for Barbara and Betty.'

'B,' said Kaarina. 'What an odd thing.'

'I'm called Neville, now. "Cocky" is just my RAF nickname.'

'Neville,' said Kaarina. 'You told me at the concert.'

'Our mother, Edith, or Edie, brought us up on her own, really. Our father, Iggy, he was sometimes around, mostly not, and we preferred it when he wasn't. He was often away at sea. We hated him.'

'Hated him?' said Kaarina.

'He was violent. Free with his fists. A big, strong man, who thought nothing of hitting his wife and his children. He was a drunkard, too, and a womaniser. He couldn't stand it if any of his children were better educated than he was. He did his absolute best to wreck all of our chances.'

'That is awful,' said Kaarina.

'It was. Now, our mother lives in her own flat in Bristol. It was my idea – give her somewhere safe to live, where Iggy doesn't know where she is. I usually pay her rent, and I'm worried about that. I can't pay it while I am here. I hope that my brothers and sisters have kept up with her rent.'

Kaarina was sympathetic, and a little shocked, but she was also trying to establish his identity beyond any doubt, before committing herself fully to helping him escape.

'You joined the British Army?'

'No. The Royal Air Force. In 1946, after the war. I've served in Singapore and Northern Germany, and I would still be there if I wasn't in the deepest, iciest wastes of Russia.'

'Who else knows you're here? Your family?'

'Not my family. My boss, Eric, knows I'm here. He's the one I talk with over the radio.'

'Is Eric trying to help you escape?'

'Not really. He's concerned that I stay safe but, to be honest, I am more valuable to him here, in the middle of the enemy camp, than I am in Germany.'

'Phone calls are monitored from here, aren't they? I used the club telephone to speak with Eino yesterday; there was a click before and after our conversation.'

'Yes, all communications are monitored closely. It's a prison camp, Kaarina, a prison with an orchestra, a clubhouse and an ice rink.'

'Neville, I am going to trust you are who you say you are. That you came to be here in the way you told us about in the boiler room. That you are being made to work on something, under duress, and you're doing all you can to frustrate that process.'

'Yes,' said Cocky.

'How long will it take to pass on your knowledge to the Russian guy?' she asked.

'Two weeks, if we are being efficient. Which we're not.'

'Could you stretch it out to two months?'

'I can try. Helmut, my boss here, is alert to time-wasting tactics. I would have to hold my nerve.'

'I heard from the people I'm staying with that Helmut's family is being made to move out of their house,' said Kaarina. 'They say that Helmut could be replaced.'

'It sounds plausible,' replied Cocky. 'It might explain why he's short-tempered ... and ... he's getting through the Russian vodka. Does Irmgard know?'

'I don't know, and I can't ask. Cocky, do you trust me enough to do as I say?' Kaarina looked at him, her green eyes sparkling and her expression serious. A stray strand of bright hair escaped from behind her ear, and Cocky resisted the impulse to tuck it back in.

Cocky knew he would do whatever it was she wanted. 'Yes,' he said.

'Find someone on the island you can buy a violin from, without arousing suspicion. Can you do that? It has to be full-sized, not child-sized.'

'Yes.'

'And, grow your hair.'

'Yes, I will.'

'And you have to give up your radio. It's too dangerous to have it, to use it. You'll be discovered before long, Helmut might tell someone, or make the connection – and that will be that.'

'Hmm, that's more difficult.'

'Cocky! You have to! Give it up, give it to me, and I'll get rid of it!'

'Yes, yes, alright, alright, I will.'

'You need a Finnish name for your visa. Oh, and another thing, keep spending time with the four of us. Get people used to there being five of us. Especially the guards on the gate. We want them to believe we are five, that we're a quintet.'

'Yes, I'll somehow manage that,' said Cocky. However terrible it is to have to hang around with you and the others, I'll put up with it for your sake!'

Kaarina laughed and punched him hard on the arm. 'Your British sarcasm! We like having you around, Cocky. It helps pass the time.'

'Helps pass the time?' It was his turn to punch her, gently, on her arm.

'We'll get you out of here, Cocky, I promise,' said Kaarina.

'Don't take any risks on my account. We've got to be so careful.'

'We are being careful. It's you who still has that bloody radio.'

'I'll bring it round tonight, and you can do what you like with it. Destroy it, if you have to.'

Later that evening, Cocky knocked on Kaarina's door holding a cake tin. 'It's in here, the radio,' he whispered.

'Ooh, a cake!' said Kaarina, for the benefit of her host family. 'Please say thank you to Irmgard.'

She shut the door, but Cocky stayed at the place where Kaarina's face had been. He didn't want to go back to his rooms. Even with Bjorn there, he would be lonely.

Kaarina opened the door again. 'Let's take this cake round to Pekka's,' she declared. 'I'll get my coat.'

They didn't go to the house where Pekka was staying. They walked deep into the woods and dug a hole with a kitchen spoon. They buried the tin as far down as they could dig – the tin that contained Cocky's homemade radio, that had never quite got as far as having its own case.

'It's too dangerous to have this in our possession,' whispered Kaarina, as if the trees were hiding malevolent ears. The only light was that of the Moon, reflecting from the surface of the snow.

'You're right. I've been stupid to hang on to it.'

'From now on, you are part of "Strava Quintet". That is your job; that is your hope. We wait for the thaw. You leave with us.'

'A dog could sniff this out,' said Cocky, looking at the disturbed snow and soil.

'No one will even be looking here,' she assured him.

'Our footprints in the snow ...'

'It's snowing, it will cover them,' she countered.

'There's someone watching us,' said Cocky. An unmistakable human-sized rustle in the undergrowth proved this wasn't just paranoia. They straightened up guiltily.

'Kiss me,' demanded Kaarina, setting her back against the trunk of a tree.

'What?'

'You heard me. Do it now,' she said. Cocky moved his cold face nearer to hers, until he could feel the warmth coming from her skin. He put his hands either side of her, leaning against the tree rather than against her, and found her lips inexpertly with his own. Kaarina dictated the length of the kiss, keeping it going until their observer's footsteps receded.

'We need to know who it is,' she said urgently. 'If it's Helmut, you are in serious danger.'

They watched the departing shape until they recognised him at the same moment: Miko!

'That's good news for you and bad news for me,' said Kaarina. 'Eino …'

'You kissed me as cover for burying the radio?'

'To protect you, yes. A reason why we are in a forest together at night. For Helmut, that would be our cover story. Now I need to catch up with Miko and tell him that what he saw wasn't what it looked like. It wasn't real.'

On Cocky's limited experience of these matters, the kiss had felt utterly genuine. To be told it was all a ruse in case Helmut had followed them to the woods sent him reeling with disappointment. He made his way back to his rooms.

'You've been out?' said Bjorn.

'Yes, I've got extremely cold. Let's make some tea in the kitchen.'

'We can't go downstairs. Helmut and Irmgard are arguing in there. Helmut's had a lot to drink and he's shouting a lot. I think we should stay up here.'

Bjorn looked at Cocky, concerned about his blue lips and shivering body. 'Look, you have the bed, have all the covers, get yourself warm.'

Cocky got into the bed. His head was swirling with images of dark woods, mischievous tree sprites that looked like Kaarina. He fell down a bottomless well, spinning and whirling, never reaching the cooling water that must be at the bottom. Rockets were flying up from the ground outside the window, and a big party of workers from the institute were outside, holding lanterns and following the rapidly disappearing footprints that would lead them through the snow to the disturbed earth over the buried radio. Except it wasn't a radio grave – it was his little brother Bernard's grave, and Bernard climbed out of it, snowy leaves in his hair, dragging a scooter and brushing soil from his dirty knees.

Another night, he sensed Kaarina there beside him. He couldn't open his eyes but he knew she was there, on the other side of the bed, asleep with her hair splayed out across the pillow. She was close and she was so beautiful. He wanted to see her so much, but someone had sewn his eyelids closed, and he cried hot tears of frustration.

He was delirious for a month, weak and ill for over six weeks. Bogdana said it was glandular fever. When he finally got up, and made his way to the window on shaky legs, the world had changed from white to brown.

Escape to Finland, 1953

The important-looking black car, a Russian GAZ-M20 Podeba with rugged snow tyres, had been waiting by the gate for nearly an hour, the khaki-uniformed driver visibly impatient to be off. A ten-hour drive was a major undertaking, starting with the deep-frozen marked vehicle track across to Ostashkov, then along slimy, pitted roads with potential for mudslides. The driver was irritated to be kept waiting, particularly because the person who had arranged for his services, a German called Gröttrup, was nowhere to be seen.

For some inexplicable reason too, there was a group of children and mothers around the car, clutching bits of bright paper and jumping up and down with excitement. This was supposed to be a highly sensitive military establishment. It was on the 'top security' list, and yet it wasn't the relentlessly grim type of place he was expecting. The people living here seemed to ignore the barbed wire fences and armed guards. It could be a neat village anywhere in Europe.

The carnival mood surged upwards as five young adults emerged from a stone-built house. They were holding musical instruments of various sizes. Long-haired and bohemian-looking, all five wore woollen hats with earflaps, in a variety of designs and colours.

The Finnish quartet had had three months to work out every detail of how to get Cocky into the car unnoticed. Bjorn had run him a bath and helped him dress in a set of Finnish clothes. They were too big, especially now Cocky was thin after his long illness. At least his hair looked the part, properly long and straggly. Even a beard of sorts had sprouted. Cocky's new Finnish name, Paavo Kekkonen, had been practised until it was perfect, and his travel visa, made so painstakingly by Bjorn, looked absolutely genuine. It had been folded, fingered and scuffed.

But no one had foreseen the crowd around the car. It put an unbearable focus on the moment of escape, just as they needed to glide away unnoticed. Anyone, child or adult, could pipe up at any moment to declare that surely there were only four people in the quartet, not five.

A few minutes before the five of them set out from the house, Cocky had taken some persuading that his disguise was sufficient to protect him from detection – detection that could result in his untimely disappearance or even death.

Bjorn had taken control. 'You will have to walk outside with us, and get in the car as if you have every right to do so. We've been talking about our fifth member, Paavo, being ill, so that is the idea we've sown around the community, and I hope it's taken root. Understand?'

'Yes, but Helmut Gröttrup will definitely recognise me, will definitely stop me.'

'Helmut is taken care of,' said Irmgard briskly. 'He will not be there. Believe me, trust me, he will not stop you leaving.'

Irmgard handed him a violin in a dusty brown case that she had stowed in a corner of the sitting room. 'Don't forget this,' she said, pushing the violin towards him. 'Dear Cocky.' She kissed him on the forehead.

'Helmut? Where?' asked Cocky.

'He's asleep in the pantry. I've locked him in. He's out cold. Please don't worry about Helmut,' said Irmgard with an air of total normality.

'Move it!' shouted Pekka from downstairs. 'We need to go, now!'

Kaarina and Miko reached the car first, and Kaarina took the front passenger seat, presumably to distract the driver while the others got in. It worked. She swished her hair and looked up through her lashes at the hapless middle-aged driver, while Bjorn, Pekka and Cocky climbed onto the back seat. As planned, Cocky immediately dropped his reindeer hat in the footwell, and bent double to pick it up, while Pekka covered him with his coat, and braced his arm heavily on top.

A small boy knocked on the passenger window. Kaarina opened it a crack, and through the couple of inches' gap, the boy tried to pass her a green and pink cardboard crown he had made. It wouldn't fit through, so Kaarina had to slide the window down further. She put the crown on, and the boy was delighted. Emboldened, he said proudly, 'We made a goodbye song for you.'

'How lovely – do sing it!' feigned Kaarina.

Stress levels inside the car were sky-high, masked by rictus grins and stiff waves of thanks to the children. Cocky was nearly passing out; his lungs had no air in them and he was seeing stars. The children shuffled around forming a line, getting their exact, practised positions right for the song. Then they moved around again to arrange themselves in height order down the side of the car. Finally, the song started. The tune was the German lullaby *'Gute Nacht mein Schatz'* but the children had used English for the words.

Goodbye to Strava
Goodbye Princess.
We won't forget you,
The fun you brought us –

Goodbye, musicians
You made us glad.
Thanks for the music,
Now we are sad.
Good-bye; good-bye; good-bye.

'Goodbye!' shouted Kaarina, and the three men waved and smiled, as more and more people pushed their cards and crowns in through the window of the moving car.

Cocky felt the pressure of Pekka's arm lift off his back, and he straightened up, clutching the reindeer hat with the earflaps as if he had only just managed to retrieve it.

'Found it,' he rasped in English.

'Well done, PAAVO,' said Miko, reminding him of his new name and nationality.

But the driver wasn't listening. He was watching Kaarina beside him, who was twinkling happily about the children and their song.

'What a lovely song from the children,' she enthused in Finnish. 'I've so enjoyed all the time we spent at their club-school, doing all sorts of musical stuff with them.'

'There are some budding musicians amongst them,' said Bjorn.

'String players,' said Miko

'Percussionists,' said Pekka

'Songwriters …' said Bjorn.

'Paavo' chuckled loudly as the others made this list, thinking it must be some kind of joke. The others jabbed him in the ribs. Kaarina had to work extra hard to keep the driver's attention on herself.

'Just be the quiet one,' hissed Bjorn to him in the back of the car. 'There's always a quiet one in every group. And if anyone questions why you're so quiet, you've been ill.'

Cocky nodded but longed to ask the questions that had been forming in his mind. Questions about what had happened in the months since they had buried his homemade radio in the cake tin in the forest. The night Kaarina had kissed him. The night he fell ill.

How had they knocked Helmut out? What was he likely to do when Irmgard let him out of the pantry? How long would it take for him to discover he had gone?

How had Bjorn done such an incredible job of forging his visa? Did the photography shop in Ostashkov have the negatives? Who had them now?

Had anyone found his buried radio?

Was Kuznetsov still around? Had he been able to make sense of his drawings? Perhaps he had returned to his laboratory empty-handed.

Had anyone heard news of Walter?

Were Helmut and Irmgard really going back to Germany at the end of the year?

Where had they found the violin? Presumably, one of the island's orchestra members had sold it to them, but were they in on the escape plan, and if not, would they say anything?

Had Kaarina really lain down on the bed beside him when he was too ill to open his eyes?

***

A few hours into the journey, the driver stopped at a small town café. The café owner was keen to serve them his daily special – cabbage soup – but they could smell sausage stew with dumplings, and insisted on having that, despite the owner's reluctance to give it to them.

While they were eating, a group of Russian workers came in, their boots dripping with mud, expecting the sausage stew that had been prepared for their midday meal. They were the regulars, but this group of six, all foreigners except for the driver, had all but finished off their precious stew. A lot of disgruntled noises and glances were coming their way, as well as a fair amount of guessing which nationality they were. The Russian driver kept up a bit of banter with them to deflect their annoyance over the stew, and in the process, he told them that the group were Finnish.

As soon as the last trace of stew was gone from their plates, the group got back into the car and the driver, who was still inside the café settling the bill, stayed a few more minutes exchanging pleasantries with the owner. Miko got jumpy about these few minutes. No one really knew whether the driver was likely to shop them to the KGB. He might even be a KGB member. Miko wanted everyone to run from the car, but there was no time. The driver came back, whistling, and set off again.

Twenty minutes later, a distinctive KGB car sped past on the outside of their car, moved them into the side, and stopped in front. Two KGB officers stepped out, and motioned to Kaarina to wind down the window.

'Papers!' said one. All six visas were collected and handed out to them and were scrutinised for what seemed like a very long time. The officers matched photos to faces, pausing just a little too long on Kaarina's.

The questions began, all directed to the driver. Where did he pick up the group? Where was he driving them to? What nationality were they, and what was their business in Russia? The boot was opened to verify that they were musicians, and all the instruments were counted out and back in again.

The officers were making moves to get back into their own car, when one of them said: 'Kekkonen. I hope you're not kidnapping the Finnish prime minister.'

Miko managed a polite, nervous laugh. No one ever expected a joke from a KGB officer, and it was very difficult to produce a response, especially as his comment showed up another flaw in their plan – Bjorn had picked a memorable surname for Paavo: it was the Finnish prime minister's name.

As the KGB car drove off, Bjorn melted in embarrassment on the back seat. 'I am so stupid, so stupid, so totally stupid, he said. Kekkonen. Kekkonen. Of all names. I was just thinking Cocky – Kekkonen.'

'It can't be helped. Shut up, Bjorn,' said Miko.

'Leave it, Bjorn,' said Pekka.

'How could I have been so ...' continued Bjorn.

'MUD, MUD, glorious MUD,' sang Kaarina loudly in English. It was one of the songs they had taught the Gorodomlya children.

'Nothing quite like it for cooling the blood,' roared Miko and Pekka.

'So follow me, follow, down to the hollow,

'And there let us wallow, in glorious MUD ...'

After a few repetitions of the rousing song, which someone had mentioned was about a hippopotamus in love, complete with furious glances towards Bjorn, he was finally able to recover his composure, and stop repeatedly admitting that he had forged Cocky's visa.

'Aaah, the car, it is covered in MUD,' said the driver happily in Russian, except for the one English word. Kaarina praised him for his correct pronunciation of 'mud', and he beamed at her with delight. He was enjoying this journey after all.

Cocky couldn't cope anymore with the strain of pretending to be not just Finnish, but also a musician. He thought it best to sidestep the difficulties by feigning sleep, even though the adrenalin from the KGB incident was still coursing through his veins.

Vaalimaa Border Control

The others chatted desultorily through the afternoon. The day appeared uniformly grey to Cocky as he gazed out of the window, simultaneously bored and in a state of high alert. A line of lights and many coloured signs appeared like stars in the distance, and then the Russian side of the Finnish border was near. A long, low and well-lit building stretched alongside the checkpoint. Only one other car was in front of them, and there were a lot of border guards for the small number of people and vehicles they had to check.

'Good luck for us,' said the driver. 'Usually, Vaalimaa is a very busy checkpoint, but we will get through fast. You'll be home in no time.'

Electricity flowed from person to person inside the car. Everyone focused on relaxing, being normal, not giving away the smallest whiff of fear. Paavo yawned and stretched too theatrically, and put on his reindeer hat. The guard motioned to them to get out of the car and take their visas into the building. Two dog handlers approached and set the dogs to work sniffing out anything the vehicle might be carrying. When the group got back from the visa check, the boot was open and the guards were lifting out the instruments.

'Open them. All. Open them all,' said the guard.

Paavo watched how Bjorn opened his violin case, and he did the same. Kaarina opened her cello case, and the cello itself was subjected to a prolonged scrutiny, with a torch being shone into the sound holes. Every compartment of the violin and viola cases was searched, and the music folder was tipped out so they had to put all the sheets back in again.

The driver was being questioned again about where he had come from. The guards believed that the group was Finnish, but not that they had come from Gorodomlya.

'Gorodomlya is closed,' said the guard. 'It's on my list of closed settlements. Any foreigners coming from those places is highly suspicious. They are perhaps not musicians. Who ordered the car?'

'Helmut Gröttrup,' said the driver. 'It's an official car. Here are the papers relating to it.'

The guards scrutinised the papers, which were genuine.

The driver volunteered some more information, because they had not yet been told to proceed. 'They are professional musicians. They've been singing in the car.'

'What were they singing? Opera?' asked the guard.

'No. They were singing ... a song.'

'Which song were they singing?'

'MUD, MUD, der der der MUD' tried the driver nervously, still pronouncing the word perfectly.

'That is a children's song. Anyone could sing it. It is suspicious, that these so-called *musicians* are leaving a closed military facility.'

The group were then questioned for over an hour about why they had been invited to Gorodomlya, and who had invited them. Names were given to the guards – Gröttrup, Korolev – and telephone numbers noted. Kaarina took the lead on giving them their answers, passing them the telephone number of the clubhouse, rather than the Gröttrups' house. She knew no one would be answering the phone at the club until Monday, and she was playing for time.

'Can we have some water? Paavo has been very ill,' said Bjorn, getting in with a reason for Paavo's silence. Water was brought, and there was a slight relaxing of the tension, as they waited for a guard to corroborate their story by telephoning the empty clubhouse. The guards had all the time in the world. The number for the club was ringing out.

The guards' next telephone call – to N11-88, Korolev's institute in Podlipki – was finally answered. The guards asked him whether there were Finnish workers on Gorodomlya. The speaker told them that as far as he knew, the workers on the island were German, apart from increasing numbers of Russian technicians following the death of Stalin a few weeks before. When the guards pressed him for knowledge of Finns, he insisted the workers were all German or Russian, but there had been a concert before Christmas where Finns may have performed.

Mitya sat in the silent, after-hours institute office, running over the border guards' phone call in his mind. The guards hadn't mentioned a British person, and he hadn't mentioned one to them, but he had a hunch that the incident had something to do with Cocky, the British scientist he had first met in Ried im Innkreis, then on the Brocken. When it came down to it, Mitya did not want to incriminate Cocky. He knew he was the only British scientist on Gorodomlya, because he had played a role in taking him there, but he felt a jolt of excitement that Cocky may be attempting escape. Years later, Mitya was able to tell Cocky that he felt pleased to have helped him, by not mentioning a British escapee to the border guards.

Back in Vaalimaa at the Finnish border: 'What about a concert?' said a guard in a bored tone. 'To pass the time?'

'Prove that you really are musicians,' said another, stubbing out his cigarette.

'We're tired,' said Miko.

'Paavo has been ill,' said Bjorn.

Kaarina shot him a look that warned him to stop mentioning Paavo.

'Come, come, a concert, a concert,' said the guard jovially, with an edge of menace.

Everyone looked at Kaarina for direction.

'We don't need sheet music,' she said. 'We can do the *Quartettino* fast movement from memory, boys. Can't we?'

Cocky froze. He remembered the fast movement from listening to his radio. It was fiendish, so fast and wild and dissonant and ... out of control. And he had never picked up a violin before. He had to hold his nerve. Bjorn positioned himself so that Cocky could copy his moves exactly. Slowly, Bjorn undid his own violin case, found his bow, and screwed the little end bit so that the horsehair was taut. Cocky did the same. So far, so mechanical. Then Bjorn reached into a pocket for a shoulder pad, fixing it in place in a practised way that Cocky couldn't follow. Cocky located his own pad, and fixed it to the wrong side of the chin rest.

'You've got a new pad,' said Bjorn, taking it from Cocky and placing it properly in place. 'Nice.'

Cocky copied Bjorn's movements, but the whole violin positioning thing was totally unnatural. He observed closely the hold, the arm and finger positions, and how the violin rested against his shoulder. He had to affect familiarity and comfort; he had to make himself relax. This was already so difficult, and they hadn't played a note yet.

'Let's tune up,' said Miko, instantly regretting it. Cocky moved the bow across the strings like Bjorn was doing. The sounds were bad, scratchy and teeth-edge tuneless. He had no idea how very out of tune the old violin actually was, only that it sounded like something dead. No one had thought to tune it when they had the chance. It was a detail they had overlooked, and now, Cocky could no more tune it than he could play it.

'Fine, let's go,' said Kaarina, launching into the piece with theatrical gusto. Cocky knew that a violin part would start later on in the piece. He waited and listened, really listened to what the others were doing, as they played furiously, wildly, dissonantly – then he joined in. With his eyes fixed on Bjorn, he stuck his chin up, gripped the bow, and dived headfirst into the worst sound-making that the group had ever known. It was dire, it was appalling, and nothing Cocky was doing was right. His untuned violin sounded cat-like, a wailing howl of awfulness. Kaarina matched the terrible sound with furious wails from her cello. Miko put in some off-key pizzicatos to deflect attention from Paavo. Bjorn banged the wooden side of his bow on the strings percussively until Paavo noticed and copied, and then the five of them found their thread again and steamed to the end.

'Bravo, bravo,' clapped the guards.

It had worked. They had got through the performance. The relief was enormous and tangible.

'Who is the composer?' asked a guard.

'Einojuhani Rautavaara,' said Paavo cockily, elated. He was a violinist! And he was speaking Finnish!

'Oh really? A modernist?'

Paavo didn't know how to say 'yes'.

'I'd like a solo,' said the other guard, grinning slowly, like a cartoon cat with a trapped mouse. 'From Paavo.'

'Yes, yes, a solo please,' said the other one, catching his meaning.

A woody, scraping, crashing sound, and Kaarina slumped to the ground, one peg of her cello smashed and rolling across the floor. Kaarina lay crumpled on the floor of the checkpoint building, her bow somewhere near her hand, her cello lying flat on its back like a whale. She looked lifeless, pale, and limp. She also somehow managed to look beautiful. The guards rushed over to her, slapping her face and checking her pulse, while Miko picked up the broken cello peg and Pekka and Bjorn put all the instruments away.

'We've not eaten since midday,' said Miko. 'She's probably fainted.'

Bjorn and Pekka observed the guards closely and would not allow them to continue checking Kaarina's vital signs. 'Please fetch food,' they told them. 'She isn't dead, she just needs to eat.'

While one guard was fetching food, the other one decided to fully unpack 'Paavo's' bag. It was the one he had brought from Britain to Germany via Gorodomlya, the one he had with him when he was bundled into the car. He hadn't unpacked it himself in all that time.

The antenna-turning mechanism, 2 feet long when folded, but 4 feet when extended, had been made in Bristol for Paddy's project. It was still there, lodged at the bottom of the bag. The guard pulled it out slowly. Kaarina, who was starting to revive, took one look at it and fainted again. Cocky looked at it in horror. This could be an oversight too far. The guards were already homing in on how he didn't fit with the group, and this potentially incriminating piece of hardware was telling a story that didn't accord.

'What is this?' asked the guard in Russian.

Cocky got up and gestured to the guards to follow him. They handcuffed him, to make sure he wouldn't try to run for it. One guard carried the device, and the other was radioing for assistance. Two more guards appeared. Cocky gestured to them to say he couldn't show them what the device was unless they undid the handcuffs. After an age, they weighed up the scrawny person in front of them and removed the handcuffs. Cocky pointed to the crop of antennas behind the checkpoint. He could see that they were fixed antennas, capable of picking up the strongest radio waves in the centre of the beam, but missing those on the periphery. He showed with his hands how an antenna using the device could be wired up to follow the clearest signal, turning automatically towards it.

'Where did you get it?' asked the Russian guard in English.

Cocky did his best to speak English with a Finnish accent.

'Britain. The Bristol Aeroplane Company.'

'It is a very good design. Why do you have it?'

'I am an inventor as well as a musician,' said Cocky. 'I designed it.'

'The first one in the world,' said the guard. 'Do you have the Patent?'

'No, I wanted to try it out first,' said Cocky.

'We will try it out, here,' said the guard.

'Yes, you may have it.' said Cocky.

'We are taking it! Confiscating it! It belongs to us!'

'You need to know how to fit it,' said Cocky. 'I'll draw you a diagram.'

When Kaarina and the others looked over towards the group, they could see Paavo drawing and explaining and moving the piece of metal around, showing where bolts should go and how wires should be fed through the mechanism. The guards were taking it all in, relating it to their own antennas, and looking pleased with themselves.

No food had materialised for anyone but Kaarina. Now she was alright again, the group asked the guards for permission to leave on grounds of hunger. They had to leave addresses to where they were going in Finland, and Bjorn put Paavo's down as his own.

The driver was heartily tired of all the waiting around, and he sped off with a flourish once everyone was aboard. On the Finnish side, a few hundred metres along the road, there were no hold-ups and the car swept past the much more relaxed Finnish border guards. As they emerged from the Finnish checkpoint, a figure ran towards them waving his arms madly.

'Kaarina! Kaarina!'

'Eino! Eino!'

'Miko! Bjorn! Pekka!'

'Eino!'

Eino embraced Kaarina tightly. All this time he had waited and worried. Now she was back, his world was apparently complete.

'Come, come, I've brought my car,' Eino urged them.

The driver lost no time in unloading the instruments from the boot of the Podeba and into Eino's car. He wanted to make sure that he could really leave the group here, and then he could head back into Russia. He picked up the remaining paper crowns and cards that were still in the passenger footwell, and crushed them securely into a bin.

'This is ... Paavo,' said Kaarina to Eino. 'He's with us.'

'What? Who is he?'

'Paavo,' said Cocky, holding out his hand to Eino. 'Paavo Kekkonen.'

'Please, Eino, it's cold. Let us explain who he is in the car,' said Kaarina.

Helsinki Bus Station and the KGB

Kaarina went over to the driver and gave him a big tip, in roubles. She hadn't been able to prevent him from seeing Eino, and his car, so she kissed him lightly on the cheek, then waved him goodbye.

As the five of them climbed into the car, Eino questioned Kaarina about the kiss she had given the driver. It seemed unwise to him, like a bribe.

'It's to do with who Paavo is supposed to be,' she began. 'We are pretending to be a quintet.' As Eino drove them back to Helsinki, loud, fast Finnish swirled around the inside of the car. Clearly, the events of the last months were being relayed to Eino. The tone of the conversation moved swiftly from pleasure to increasing unease, and finally to anger on Eino's part.

Cocky could only pick out bits of the heated language coming from Eino, but he appeared to think that even if Cocky was a genuine British person, who warranted all the risks the group had taken, they had by their own admission practically left a trail from Gorodomlya to their houses: the man locked in a pantry; the disappearance of the British engineer; the negatives of the forged visa; the bits of paper crown and card that probably littered the road they had taken; the Russian workers in the café; the KGB officer who had noticed the surname; the distinctly bad violin playing; and, to top it all, the existence of a highly engineered piece of military equipment – something a musician would never possess – actually in the British man's bag.

All the clues made an easy task for any member of the KGB wanting to find the fugitive. He couldn't believe they had got through the Russian checkpoint without being arrested. He was furious, too, that their bumbling scheme had another serious flaw – no one had planned what to do with Cocky once he was free. And now the man was in his car – *a wanted man* – and he fully expected the Russians to be on his tail.

Eino stopped the car on the outskirts of Helsinki. It was a bleak, empty piece of scrubland. 'Out,' he said to Cocky.

'Not here, not here, Eino,' pleaded Bjorn. 'At least take him to the bus station. He stands a chance of getting a bus to somewhere from there.'

Eino reluctantly agreed to drop Cocky off at the city's bus station. When Bjorn was taking Cocky's violin and bag out of the boot, he whispered to him: 'Stay at the bus station. Don't take a bus anywhere. I will come back for you.'

As Eino and his passengers sped off with a screech of tyres, Cocky considered his position. At this late hour, the bus station office was closed and the only place he could find to sit was in the bus shelters. He peered into one, but a dark shape was stretched out on the bench – a down-and-out, no doubt. How disturbing. He might be violent, drunk, or at least angry at being woken up. He looked around the other shelters. Some had occupants, others not. He thought longingly of his big bed on Gorodomlya. His stomach growled with hunger. The sudden loss of the group's friendship, which had kept him going for so long, seemed to punch him in the face.

On the night air, the smell of fried fish came wafting over. It must be a figment of his imagination, he surmised; there were no houses, no cafés around. He must be hallucinating. He took a few paces in the direction of the smell, before spotting a tiny orange light a distance away. It was flickering, but definitely there, so he walked towards it, carrying his violin and bag, with no other plan.

As he approached the orange light, he saw it was a campfire. He could make out the shapes of people gathered around it, and caught the low hum of voices. Getting closer, he could see a kettle on the fire, and a black pan, full of gutted fish. A pile of fish heads and entrails was on the ground, and a few cats milled around it, already sated.

The conversation stopped as Cocky stepped into the circle. A few hands reached for concealed knives. Abrupt questions came at him from worn and grizzled faces eerily lit up by the firelight. He couldn't understand them, so he said 'British. Cocky' over and over, until the ringleaders of the group had absorbed the message of his nationality and name, and had assessed him as no threat to them. One of them reached for an enormous basket of fresh fish. The lettering on the side of it showed that this was the catch of a trawler, probably illegally removed from a harbour warehouse. More of the down-and-outs Cocky had seen asleep in the bus shelters were making their way to the campfire too. The first man, the same one who had perturbed him on his arrival at the bus station, sat down next to him and said something about the violin.

It was imperative to trust these people, these homeless people, these fish thieves. He thought he knew their type, due to all the unexamined prejudices swirling around his overwrought brain. Violent – drunk – thieves. He made a determined effort to suspend his fears and look into the face of the man who had asked about the violin. The man stared back, a question in his eyes. The violin, Cocky now thought, was probably valuable. The man might wait until he had a chance to steal it, but if Cocky reacted in that way, the man could respond badly. Then he wouldn't get any fried fish, and he might even be set upon and killed. He tried again to return to the imperative of trust. It was so difficult to still his fears.

Someone passed round a box with bread in it – it was similarly stolen from somewhere, and no longer fresh, but it was glorious to Cocky, and he broke off a big hunk of it. The man took it and broke off some for himself, leaving dirty fingermarks on the loaf. They chewed for a while, and then the fried fish was passed around. Cocky had never tasted anything so delicious – black and crispy on the outside, hot, fresh and flavoursome on the inside. Tin mugs of black tea were passed around, too, and these had to be shared because there were not enough of them. The fish kept on coming – there was no shortage. For a long while, people just ate in silence.

The man was pointing to the violin again. He mimed a playing motion, and asked in gestures whether Cocky would play for them. Cocky shook his head and pushed the violin towards him. Slowly, the man opened the case and lifted out the violin, tenderly and with skill. He found the shoulder pad and bow, but only unscrewed the bow a little to relax it, and ignored the shoulder pad. He plucked a string, grimaced, and then tuned it. He tuned all four strings. He stood up. He pulled the bow across the strings in a crashing chord, and then launched into Tchaikovsky's *Canzonetta* – a sad, Hasidic melody. A woman in a ragged dress stood up to dance, moving sinuously to the melancholy tune. Cocky was amazed at the sound of the violin. It was sweet and rough at the same time. It sang and sighed. It moved him profoundly.

A car appeared in the distance and moved closer, but the road was a few hundred metres away, and the driver drove by. Cocky thought of Bjorn, who had said he was coming back for him. The fiddle player was in full flow, tune after tune coming from his fingers on the instrument, and people were clapping and dancing, and singing along. Cocky stood up to see where the car had gone. When he sat down again, he saw that all of the men had their hands on the knives in their pockets. They were having trust issues, too.

Would Bjorn come back? It must have been Bjorn in the car. He wondered whether to walk back down to the darkness of the bus station. Surely, Bjorn would look for him, would find him here at the fire?

He noticed something under a cover near the empty basket of fish. It was a box-shaped radio, very old and battered. It had a carved wooden case, so must have been a good one once. He pulled it out and took it back to his place to see it better in the firelight. Two of the dancers, a man and a woman, moved to protect their radio.

'Is it working?' he asked them, as he turned the knob and adjusted the broken antenna. 'No,' they gestured. 'It stopped working long ago.'

Just holding the radio was calming to Cocky. He wanted to see inside it, the wires and transistors and diodes; he wanted to connect with its innards and feel like he was on home ground. He needed a screwdriver, and he needed to keep a lookout for Bjorn, who would surely find him soon.

'Screwdriver?' he mimed to the group.

The men with knives all moved at the same time to pull them out of their pockets. The ring of blades was impressive – all sizes, lengths and degrees of sharpness. They used them all the time, for gutting fish, for living. For protecting themselves. Cocky walked around the circle and selected a long, thin blade that would work as an improvised screwdriver. The rest of the blades were put away, and Cocky set to work undoing the radio casing.

'You waiting for someone?' asked the woman of the radio couple in heavily accented English. 'Someone in a car?'

'Yes,' said Cocky. 'A friend.'

'Not police?'

'Not police,' confirmed Cocky.

He saw at once where wires had come loose from their contact points. He pared back the insulation and reinserted them, screwing down the hold points firmly. 'You need a battery,' he said. 'Then it will work.'

'We have a battery,' said the man of the couple, and he disappeared behind a pile of household junk, which Cocky now saw was in fact a makeshift shelter.

The man came back with a tin – a tin in which he collected things that people might find or steal, that might come in useful. Every household has such a tin, he thought. Including this one. He cleaned the battery and put it in the radio, screwing the casing back down and even cleaning the outside of it. The fiddle player stopped playing so that people could hear if the radio worked. A blare of rock and roll came out of it, fast, heavy, and rhythmic, and there was a ripple of appreciation around the group.

Cocky grinned at them and they grinned back at him. It was an unforgettable moment. 'Have it,' he said to the fiddle player. 'Have the violin.'

'No, I can't pay. I can't pay you for it. No money,' said the man who had played so amazingly well. 'It was enough, enough for me just to play it again, this one time. Thank you, British Cocky.'

The woman pulled Cocky's sleeve to bring his attention to the same car that had passed by before. The car had stopped a way off, so Cocky left his bag and violin with the group, and went to talk to the driver. It was Bjorn, but that wasn't immediately obvious because he had disguised himself with dark glasses, and removed his woolly hat.

'I didn't think you'd be with those people,' said Bjorn. 'They're dangerous.'

'No, they're not dangerous. They look scary but they are friendly.'

'Cocky, listen. Get your bag. I can't take you home with me, Eino is right, because the border guards have my address. My mother, who I live with, knows nothing about all this, about you, and that's the way it needs to stay. But I've found you a place to sleep for tonight. With a friend of mine. You will have to leave in the morning.'

'I need to get to the British Embassy tomorrow,' said Cocky. 'To get a passport, and money, to get a plane ticket home. To the UK.'

'Get your bag, Cocky, hurry up, and don't forget the violin. You'll need to sell it. You can only stay with my friend for one night; you'll need the violin money to live on for a few days.'

Cocky walked back over to the campfire, which was dying now. His bag and the violin were set neatly side by side on the grass. He picked them up and looked around the group. There was nothing but goodwill and openness radiating from the faces around the circle. His heart was full and he felt the urge to say something meaningful, but only managed 'Goodbye'.

'Go well,' said the woman who had earlier spoken to him in English.

The others picked up on the phrase and repeated it: 'Go well – go well – go well.'

The next day at the British Embassy in Helsinki, Cocky was able to start the process of obtaining a passport, and the tricky issue of getting his salary money wired over from the RAF, which he would need for his flight ticket. It was all going to take time, and he needed to find a guesthouse for the duration, as he had nowhere else to stay. He resolved to sell the violin.

There was a moment in the music shop in the backstreets of Helsinki town, just before he opened the violin case, when the thought occurred to him that the violin might not be inside. The homeless people at the campfire could easily have removed it and weighted the case with something else. They'd had the opportunity. The shop owner had the case on his counter, his monocle fixed in his eye ready to examine it, and was gently lifting the lid. To Cocky's relief, the violin was intact. Moreover, it had been wrapped in a soft, coloured scarf, which he recognised as belonging to the woman at the fire, the one whose radio he had mended. Her kindness, and the memory of the warmth of the people at the campfire, hit him forcefully. Tears started to form, threatening to trickle down his face.

The music shop owner had noticed Cocky's demeanour, despite his focus being on the violin. 'You know, selling a violin is an emotional thing, a sad thing,' he said. 'It's not just an instrument. It's a friend.'

'Yes, it means a lot to me,' said Cocky. 'But I have to sell it.'

'Would you like one last chance to play it? I'm not in any hurry today,' said the shop owner kindly.

'Oh no, thank you. I'm not very good,' said Cocky, terrified at the thought of pretending to play it again.

The owner chuckled. 'No one thinks they are any good. These soloists are so self-critical. They can stand on the stage, delighting the audience for hours, with their wonderful playing, their depth of feeling, their sheer brilliance and the beauty of their sound ... then say that they're *not very good* because they fluffed a phrase or two, something that no one in the audience would even notice.'

Cocky stood there, not knowing how to respond. 'I'm really not a soloist,' he said humbly.

'You've been one, though, at some time? When was your last concert?'

'Um, the Queen Elizabeth Concert Hall, in London, shortly before it was bombed,' said Cocky.

'Ah, yes, the London Philharmonic. A bad business, such a bad business. Such a beautiful, perfect concert hall, destroyed. So you haven't played solo since then?'

'No,' said Cocky miserably. 'Can you tell me what the violin is worth now?'

'Well, it is in very good condition. I see you've looked after it,' he said, observing the soft scarf that was wrapped around it. And it's in perfect tune, after all this time. The pegs and tuners will need a little attention, and maybe the top two strings are a little worn, but otherwise, it's perfect. As you know, it's an early model, by Theodor Berger. The label inside has the words 'Berger, St Petersburg', and a date, '1897'.

'Yes. An early Theodore Berger,' repeated Cocky.

'Worth today, seventy-thousand Krona,' said the owner. Would you like a cheque, or cash?'

Cocky opted for cash, because he had no bank account. He signed the receipt for the big stack of cash, using his real name, and the paperwork to say that the violin was his to sell. He wasn't entirely sure if the violin, which one of the Finns must have bought for him from a German player during his illness, was indeed his to sell, but he would find out who had bought it. He had no concept of Finnish Krona, but he guessed from the owner's reaction to the violin that this was a lot of money.

The owner offered him the soft scarf back, as a memento of the violin. Cocky took it, and a plan started to form in his mind. First, he found a guesthouse on the edge of town, and paid in advance in cash for a week's stay. That might be long enough for his new passport to come through. Then he went back to Bjorn's friend's house, where he had stayed the night before, and left his guesthouse address for Bjorn.

Bjorn came around that evening. They sat in the small sitting room of the clean, pleasant place, and Cocky had just asked who had paid for the violin they gave him, when Bjorn said, 'The Finnish police came to my house yesterday, when I was out picking you up from the bus station. My mother confirmed to them that I had returned home yesterday night, but that she hadn't seen a second man, calling himself Paavo Kekkonen, posing as a Finn, but probably British.'

'The Finnish police?'

'Yes. The Russian police must have passed the arrest warrant to the Finnish police.'

'Arrest warrant. Oh no.'

'Cocky, listen. The Finnish police made the visit to my mother's address. But they spent most of the time on her doorstep admiring her geraniums. They don't want to follow Russian requests for arrests and, to be honest, they would resist any form of coercion to do so. The Finnish police act in the interests of independent Finland. They called at the address they were given, you weren't there, they won't give it another thought.'

'Was your mother upset about the police calling? Did she tell them anything?'

'My mother thought they were charming, and she was out this morning tending to those geraniums as if they were her children.'

Cocky was relieved, and the conversation moved back to the violin.

'Bjorn, did you buy the violin from someone on Gorodomlya?'

'We clubbed together for it,' said Bjorn. 'It wasn't much. It belonged to someone's grandfather, who had died, and no one else in their family played. We paid ... I think altogether we paid the equivalent in roubles of about five-thousand Krona.'

'Today I got seventy-thousand for it,' said Cocky.

'Wow,' said Bjorn. 'We never even looked at it. We didn't tune it up, either – as you found out. We just assumed it was a standard orchestral violin, not some kind of collector's piece. Only a virtuoso would really be in the market for something like that.'

'I want you and the others to share the profits. And I want to give some of it to the homeless people at the bus station.'

'Alright. I'll tell the others. I think we will all be pleased. It will make up for the earnings we lost on Gorodomlya, and pay for Kaarina's cello to be fixed. The pegs got broken; they will need to be remade.'

Cocky had forgotten about the broken cello, and he only now realised what a sacrifice Kaarina had made by pretending to faint just at that moment. Her precious cello. 'How is Kaarina?' he asked.

'She's fine,' said Bjorn guardedly. 'Happy to be home.'

'Eino was …' began Cocky, 'angry?'

'He was angry with all of us. For sheltering you, believing you, putting ourselves at risk for you, and he was angry about all the blunders. We made a lot of blunders, maybe because we're not naturally deceitful people. And we're not trained spies. But any one of those mistakes could have led to us all being imprisoned, even killed. We should have been more careful about things like the photography shop in Ostashkov, which still has the negatives of your forged visa. I should have thought more about the surname, picked a more common one.'

'And I should have remembered I still had a huge piece of antenna-turning equipment in my bag,' said Cocky ruefully. 'I told them it wasn't patented, but it will be by now, probably by Paddy.'

'Paddy?' queried Bjorn.

'Never mind,' said Cocky.

'I don't know which of our mistakes was the worst,' said Bjorn. 'But we got through. We are home, and you will be soon. You can buy an air ticket to London as soon as your passport arrives.'

That night, Bjorn drove Cocky back to the place near the bus station where the campfire had been. Cocky had wrapped up a wodge of cash in the woman's scarf, an amount that would enable them not to have to steal food. But the campfire was cold and grey, and the makeshift shelter he had seen the night before was pulled apart and scattered over the ground. The fish entrails hadn't been buried and were giving off a terrible smell. The destruction must have happened shortly after he had left, but why?

They looked at the scene for a long time, wondering where the people had gone. A soft knock came at the car window. It was the woman who had wrapped the violin in her scarf.

'There were two men, not Finnish, Russian, I think. Their motorbikes were Russian. We think KGB but the secret ones, in plain clothes. They asked if we had seen a British man. We said no. They didn't believe us. They punched us, hit us with clubs, and destroyed our shelter. We threw a bucket of fish slops over them as they left. It didn't hit them, just splashed them. We wanted to show what we thought of them.'

'The KGB,' said Bjorn. 'Oh God.'

'It's my fault,' said Cocky to the woman. 'I put you in danger by joining you.'

'No, this was not your fault,' said the woman. 'You did not do us harm.'

'I came in peace, and you accepted me in peace,' said Cocky, omitting to mention all the evidence of distrust on both sides. 'Here, take your scarf back. Thank you for wrapping the violin in it. There is money in it that will help you. I am so sorry about what the KGB did to you. Thank you for telling them that you had not seen me. I wish you well.'

'Goodbye, go well,' said the woman.

They drove off, leaving her in the middle of the road. She still hadn't looked inside her scarf.

Bjorn was beside himself with anxiety. 'Cocky, the KGB are not people to mess with. They're right behind you. I mean, they were here, looking for you at a bloody campfire in the middle of nowhere. They will find you, it's only a matter of time. You need to get to London, now, today – don't wait for them to catch up with you.'

'I don't have my new passport yet,' said Cocky.

But there was another reason that Cocky didn't want to leave just yet, didn't want to pull RAF strings to get himself out. He hadn't seen Kaarina since she'd been driven off in Eino's car. He felt they had unfinished business, questions not asked, thank yous and goodbyes not said. Every time he changed guesthouse, he made sure Bjorn knew where he was, so that she would be able to find him.

He didn't want to leave the country until he had seen Kaarina. Even if the KGB were after him. But his friends made it clear he had to leave. Both Kaarina and Miko's houses were being watched by mysterious men, and Pekka had received silent phone calls. It was unsettling, and Eino was so concerned about Kaarina's safety that he begged Cocky to leave Helsinki.

When Cocky's passport arrived, he picked it up from the British Embassy and bought a one-way Finnair ticket to London. As he waited at the airport, he scanned up and down the concourse, hoping that Kaarina would materialise. Several women came through the doors who looked a bit like her, but when something about their face or figure betrayed the fact that they were someone else, he was disappointed.

The flight was bittersweet. He was finally free of fear, but he would never see Kaarina again.

Chapter 6

Back to the UK, 1953

Coronation Day

Cocky was making his way to the Air Ministry building in central London for a debriefing with Churchill and an invited group of GCHQ personnel. He was looking forward to Eric being there too. The sheer number of people in the streets amazed him. It was summer, but it was raining and they must be getting wet, but the crowds showed no sign of caring. It was difficult to make his way along the pavement. He drank in the cockneyfied English being spoken all around him, enjoying feeling safe in a crowd for the first time in ages. He tried to work out what all the people were doing there. He had asked a few people, but they just laughed in his face. 'Ha ha. Good one!'

Suddenly, the crowd roared as a coach and eight grey horses came into view, a golden royal coach, carrying a young, dark-haired woman dressed resplendently in white, with a velvet cloak, and a tall man by her side. 'Long live the Queen,' shouted the crowd, and some tried out her new name: 'Queen Elizabeth!' 'Queen Elizabeth!' The crowd strained forward to get a better view. Cocky felt slightly guilty that he had one of the best views without having camped out all night or waited around for any time at all.

The coronation. Of course! He had heard about the death of Elizabeth's father, King George VI, but it was so removed from his own life that the events playing out now before him were totally unexpected. The front coach went by, followed by more splendid coaches, more royal people, regimental bands, drummer bands, exotic Commonwealth coaches, and huge processions of military men – from the army, navy and RAF. Cocky stood, mesmerised by the noise and spectacle. It was overwhelming after quiet Finland and claustrophobic Gorodomlya.

Winston Churchill, whom he thought he was going to be meeting that day, must be somewhere in the procession, as the Queen's prime minister, or with some other honorary title. Eric must have told him the wrong day for the meeting, or maybe he had noted it down wrongly. The Air Ministry wouldn't have double booked Churchill for this day of national importance. He reached the Air Ministry building and, predictably, it was closed. Eric arrived at about the same time, similarly thinking that the meeting was today, and after a hearty embrace, the two friends found their way to a backstreet restaurant and sat down eagerly to catch up on years of news.

'You got out,' laughed Eric. 'Amazing! I thought you'd been discovered. Your radio went dead. I tried to receive your signal, night after night, I never gave up. No one gave up on you'.

'The radio was buried in the woods,' grinned Cocky. 'It was too dangerous to have in my possession.'

'And what's this about escaping disguised as a Finnish rock and roll singer?'

'A member of a classical Finnish quintet,' corrected Cocky. 'A violinist. How do you know that?'

'I telephoned Herr Gröttrup, pretending to be German. I wanted any news of you. But I got his wife, Irmgard, who told me that you were no longer on Gorodomlya. And that she and Helmut were leaving too, being sent to East Germany. She told me that some people on the Island thought there were four members of the Finnish rock group, and others swore blind there were five, and she seemed to find that quite entertaining.'

'It wasn't a rock group. Did Irmgard know that the KGB found out?' asked Cocky.

'She told me the KGB had been on Gorodomlya looking for your passport and for evidence of the forged one too. They didn't find anything. Beyond that, she didn't know if you'd made it out of Russia.'

'They nearly tracked me down in Finland. The KGB. I don't think I'll ever be safe from them. They have me down as a spy, for passing secrets to the Americans. I didn't know enough for that - I can only assume that someone else must have been doing that. If anything, you and I passed a few technical details the other way. Even so, the KGB are just a step behind me all the time.

I won't be able to come back to the ASRU at Obernkirchen, my nerves are in shreds.'

'Maybe not Obernkirchen, then,' said Eric. 'It's called the 646 Signals Unit now. I've been thinking. You would be valuable to me somewhere else. RAF Gatow, Berlin.'

RAF Gatow, Berlin

The Signals Unit at RAF Gatow in West Berlin was smaller than the one at Obernkirchen, occupying just one room at the base of the control tower. Only six people could work there at any one time, although others travelled around the communist Eastern Bloc, coming back like bees returning to the hive. Its main advantage was its location – just across the border from East Germany, then known as the DDR, at that time under Soviet control.

It was 1954, and Eric had forged close working relationships with the US Air Force Office of Intelligence, and was frequently invited to lecture on intercept problems and results. At Gatow, there was none of the easy camaraderie that had oiled the wheels of effective communication at the ASRU. At Obernkirchen, listeners to Russian signals had occasionally been surprised at the end of a long shift by a hilarious blast of *The Goon Show*, courtesy of Eric. The lightness of the culture there, with its skittles tournaments and pantomimes, helped the morale of those who listened, filtered, prioritised and passed on top-secret information to GCHQ.

One consolation of working at Gatow was close liaison with photographic surveillance pilots. These airmen flew across the vast Russian continent, looking for installations and unusual movements of vehicles, building up a picture of Soviet nuclear readiness and intent. From one of these surveillance flights, photographic evidence revealed a massive, newly built city that was not on any maps. The city was far from Moscow, near to Yekaterinburg, on the far side of the Ural Mountains.

Cocky had once told Eric about the rumour of a plutonium production plant that had first emerged at the ASRU. The rumour had also been circulating while he was on Gorodomlya. Now, in Gatow, the means of investigating it were in place, and the surveillance photographs not only showed that it existed but also revealed its sheer size and scope.

'A closed-off city, thousands of miles from Moscow, and with all the identifying signs of a rudimentary plutonium plant,' said Eric to the team. 'This city will be set up to produce as much plutonium as the Soviets could ever need. It's very likely to be the fuel store for their nuclear weapons. Their name for it appears to be just a number – City 40.'

'How near are they to actually being able to use the plutonium they're making?' someone asked. 'Where are they getting the workforce from to build it, in that remote area? Do they know enough about handling nuclear fuels to do it safely?'

Finding City 40 opened up the possibility of discovering other, secret installations, hidden in mountains and underground tunnels, in forested areas and on remote islands. The photographic pilots and some ground-based agents were dispatched to find out more, following a frenzy of hunches and clues.

Cocky often wondered what had become of his work on the inertial guidance system that he had never managed to pass over to Kuznetsov. To try to find out, from Gatow he targeted the signals coming from Kuznetsov's laboratory, N11-10, working to pick up any signals or transmissions that would tell him whether the inertial system was in place. The answer came not through signals intelligence but photographic surveillance. An enormous area, about 125km long by 85km wide, was being cleared and buildings being built in Baikonur. For a long time, the purpose of the huge installation could only be guessed at, until the day when three huge antennas were erected around the site. Radio antennas were only necessary for a radio-controlled rocket, not one guided by an inertial system, which could be launched from a very small site.

Cocky was able to deduce from the radio antennas installed at Baikonur[1] that the inertial system he had worked on must have been abandoned. Over a year went by with him thinking that his work on the guidance system had thankfully come to nothing, and that Kuznetsov had not been able to improve the guidance capability of the R-7. But shortly after a trip home to the UK, Eric told Cocky that an unusual transmission had been coming in from

N11-88, the main laboratory near Moscow, and that although it was heavily encrypted, one of Cocky's previous call signs had been used repeatedly, and that the codename 'chicken stew' appeared to be attached to the transmission.

'Chicken stew,' repeated Cocky. It was dimly familiar.

'I remember a particularly good chicken stew at the Fraunhofer Institute,' mused Eric. 'Marianna made it.'

'Just what I was thinking,' said Cocky. 'I think that this signal must be from Mitya. When I was being held on the Brocken, and Mitya was part of the group of Russians holding me, he brought me some cabbage soup. When he was at the door with it, when no one else was listening, he emphasised that it wasn't *chicken stew*.'

'He was establishing a code word with you?' asked Eric.

'I didn't think so at the time. It was just a joke between us. I thought he was merely showing a little humanity.'

The next few days were spent in anticipation of messages from 'chicken stew', but a couple of months went by before a similar transmission came through. It was made during the early hours of the Russian morning, and it was very short – just Cocky's code name and the code word. Cocky had instructed the signals listeners to report any mentions of the code word to him, and when he was called in to the station that morning, it was because the transmission was still live. Cocky took over the headset and receiver, and tapped back a message to confirm his identity. After a long pause, during which Cocky felt sure that the caller had been interrupted, discovered, or prevented from continuing, the transmission resumed, and the caller set up a place, date and time for a meeting the following month, in Potsdam, East Germany.

Cocky logged all the particulars of the call and met with Eric to discuss it and decide what to do next. They chewed over the likelihood of this being a trap. Cocky presumed he was on a KGB list of wanted people. He knew that casual murder with a cover story, such as an unfortunate accident, was their operating mode.

He hadn't made any trips into the communist Eastern Bloc since returning from Gorodomlya, and he had no way of knowing if this was indeed Mitya, or someone pretending to be him. It was hard to tell if the contact was friendly or hostile. It could be that Mitya was being used as a lure. Mitya could easily be working closely with the new Russian leader, Khrushchev, to identify and prevent leaks to the Americans. On the other hand, a genuinely friendly Mitya could mean access to information of the highest quality and usefulness, working as he did at the heart of the Soviet rocketry and atomic efforts. It was agonising trying to decide what to do, how to protect himself, how to be sure of returning from the meet-up alive. He didn't sleep for a week, and Eric kept oscillating between sensing a trap and coming up with reasons why Mitya might be on their side.

One day, Eric had been awake and thinking deeply about the issue, remembering a conversation with Mitya in the kitchen garden at the Fraunhofer Institute.

'Korolev. He and Mitya were very close. Remember, Korolev thought of Mitya as the "son he never had". When Stalin "purged" Korolev and sentenced him to years of hard labour in that gold mine, he told Mitya, who in turn told me, one of the reasons for Stalin's actions was that another scientist had "denounced" Korolev for dragging his feet on developing a weapon, but also, that Korolev had upset the Kremlin by departing from the Party line. Apparently, he had dared to suggest that another use for rocket technology might be space exploration.'

'Upset them – because that wasn't the Communist Party line?' queried Cocky.

'I can only assume so,' said Eric. 'With a dictatorship, no one is allowed to express divergent views.'

'It's a more peaceful use for rocket technology,' said Cocky eventually. 'These rockets are the first artificial objects to penetrate beyond the ionosphere, into space. Why would Mitya want to talk to us, the British, about that now?'

'I imagine that things are getting critical at the Kremlin,' said Eric. 'The rocket is ready, the atomic warhead is ready, the launch site at Baikonur is nearly ready. Imagine if, all this time, Korolev has secretly been trying to avert war. Don't forget, he didn't want to be involved in making weapons in his youth; he wanted to be a peaceful scientist. Remember how jealous he was of Professor Dieminger, being able to go off and devote his life to scientific research?'

'He was definitely jealous,' conceded Cocky. 'It's certainly a possibility that Korolev doesn't want these weapons to ever be used, and that this is a way of changing their use, generating a face-saving cover story for all of this hostile activity, making out that all along, the Soviets were just interested in space exploration, not nuking other countries to smithereens.'

'Think about it,' began Eric. 'The Soviets would say that they didn't start the Cold War. When the Americans demonstrated the sheer destructive power of atomic bombs, in Hiroshima and Nagasaki, there was no way that they could afford to sit back and wait to be next. They were scared, and felt they had to get an atomic bomb as well.'

'It's gone way past the level of deterrence,' agreed Cocky. 'On both sides.'

'My hunch is', said Eric, 'that Mitya wants you to help him. Maybe Mitya too just wants to prevent war. Something that could prevent a nuclear war, at this point, would be evidence to show that the Americans are planning to be the first in space. Then the Russians could compete with that. They'd have a way out, a climb-down. The Kremlin would get their opportunity to show off their rockets, maybe even be the first into space, and gain international recognition.'

'How do we get the Americans to divert their efforts towards space exploration?' asked Cocky. 'It's hardly top of any national agenda, is it?'

'It's already starting to happen,' said Eric. 'The Americans are starting to see the potential for a space competition instead of a terrible war.'

'Let's imagine that Korolev's intention is to prevent their use for war. He is trying to divert these rockets, trying to send them up into space instead of …'

'Into America, or Europe,' agreed Cocky.

'That's why Korolev needs Mitya, who needs you – to provide evidence that the Americans are serious about space exploration,' deduced Eric. 'Then Korolev's got a reason to compete on that. If we are right about this, the Kremlin will divert all the weaponry developments to focus on space. They will want to be first in space, best rocket scientists, most prestigious.'

'You think it will be safe for me to meet with Mitya, in this café in Potsdam?' said Cocky.

'Maybe change the venue at the last minute,' said Eric. 'And I will arrange for some plain clothes protection officers to be in the café that you choose. But yes, I think you should go, armed with evidence of American commitment to a space programme.'

* * *

Cocky arrived at the café and recognised Mitya immediately. He caught his eye, and said he knew a better café along the street. Mitya showed no concern about the change of venue. He just finished his Turkish coffee and followed Cocky as he dived into a side-street coffee house and found a corner table. Mitya looked older, his hair no longer long, his acne scars no longer livid. Only his hooked nose hadn't changed.

After some chat to test the water, where anyone listening would have thought the two men were sharing a reminiscence about a particularly good chicken stew, Cocky asked Mitya about Baikonur. Mitya was quiet for a while, weighing up the consequences of his answer.

'There was a huge row,' he said eventually. 'The inertial guidance system is in place, credited to Kuznetsov. But Korolev didn't prevent all those huge radio antennas from being built on Baikonur. The size of the site was predicated on having radio control towers. Korolev was heavily criticised by the Special Committee. He should have been clear that launch pads for rockets with onboard guidance are so small they can be mobile. Hundreds of thousands of roubles, and two whole years, were wasted on the Baikonur infrastructure.'

Cocky remembered demonstrating a rudimentary gyroscopic inertial system to Korolev and Kuznetsov a few weeks before Christmas in 1952. 'Did Korolev really not know in time? I thought Kuznetsov ... Could Korolev really not have prevented an elaborate launch site being built?'

Mitya slowly chewed a pastry to allow time for his thoughts to form. 'He did know. He met with Kuznetsov at least a year before they started building.'

'Was Korolev delaying, so that things would take longer?' probed Cocky.

'He could have told them *a lot* earlier,' said Mitya. 'It would have saved *a lot* of time.' He leaned forward, his voice low and urgent. 'He ... and I ... and many people ... we are trying to prevent this madness. Since Stalin's death, we've been able to be more open about our fears over where all this is leading. Korolev never wanted to be the Chief Designer of War.'

'Eric thinks that Korolev was once punished for suggesting that rockets could be used to explore space,' said Cocky.

'Yes. Stalin considered that to be against the Kremlin's express desire for extremely powerful missiles. To keep the USSR safe from attack. He demoted Korolev. Even when he was reinstated to the same post he held before, he was, and still is, subjected to "loyalty tests".'

Cocky wondered what the 'loyalty tests' consisted of. He didn't imagine that they were pleasant. 'What if the Kremlin decided that space exploration would bring them international respect?' he asked. 'What then?'

'It would have to be Khrushchev's idea, not Korolev's, not mine, and certainly not yours,' said Mitya slowly.

'You could provide evidence that the Americans are thinking along those lines?' said Cocky. 'You could bring back from this meeting with me today evidence that the Americans are planning to be first in space with their rockets. First beyond the ionosphere. First to reach weightlessness. First to examine the Moon from close up.'

'First to show that they have the power to send a rocket a huge distance, to wherever they choose, more like,' said Mitya.

'Yes.'

'You're saying that the Kremlin could choose to use their knowledge and power for peaceful purposes,' said Mitya, 'and be celebrated for it.'

'Does the Kremlin care about its reputation with other nations?' asked Cocky incredulously.

'The USSR has never experienced being praised. It's only ever played the tough card. Seeking praise from other nations is seen as weak. They want respect. For their technological advancement,' whispered Mitya.

'It's what they do to people,' said Cocky. 'That means they don't get respected. Other nations fear them, not respect them.'

'Cocky, it's the same thing in their eyes. Fear and respect.'

'Like my father,' said Cocky.

'What?'

'Nothing. But Mitya, you're saying "them" as if that's not your opinion,' said Cocky.

'The Party Leadership', said Mitya, 'is infallible. I don't have a personal view.'

'Do you trust your leadership?'

'I've never thought about it. My job is to obey. Or get quietly disappeared. That's *my* fear.'

'In the eyes of the West, that's exactly what it is. The rule of fear – that's why the Russians aren't praised,' said Cocky. 'That's why the West finds it so hard to build trust. My colleague Walter …'

Three men at a nearby table, who were slowly eating a hearty meal, glanced over to where Cocky was raising his voice. Cocky hoped they were the protection officers Eric had promised him, not Russian spies. It was so hard to be sure.

He calmed himself down. It was essential that he played this next part of the conversation calmly and correctly. 'Eric's just got back from a four-day conference with the Americans at the Air Ministry in London,' he said, conversationally. 'I just happen to have on me some academic papers – about plans to use rockets for *space exploration*.'

Mitya didn't react. He didn't dare express too much interest, and consequently expressed none at all.

'The R-7, for example,' said Cocky, 'is an ideal vehicle for reaching outer space, maybe with instruments attached, to measure all sorts of interesting phenomena.'

'Korolev has already tried to make that case,' began Mitya. 'Last year, as I told you, he …'

'But the Kremlin didn't know, at that time, that the *Americans* were thinking along those lines. That they might be gearing up to do it *first*,' cut in Cocky urgently.

Mitya understood that Cocky had pre-empted exactly what was needed.

'Can I see? The academic papers?' asked Mitya.

'You can take them. I've got the proceedings of the whole conference back at base. These are just the papers Eric picked up in the plenaries.'

Cocky handed over the papers. Mitya glanced at them and put them in his bag.

'Cocky,' he said, 'you are still wanted by the KGB. You're still on the list of people they want to "neutralise". I won't ask you to risk meeting with me again. We can communicate over the radio, but I must stress, the sooner you can get yourself right out of Gatow, the better.'

'Wherever I go, the KGB will follow. They have sleeper cells everywhere.'

'Not all KGB are loyal to the Party,' said Mitya. 'More and more are secret dissenters – it's not as clear cut as it was. But you must take no chances. And tell Eric – he's on the list too.'

Cocky covered his shock by pulling on his gloves and hat.

They left the café, followed by the three men.

Eric assured Cocky later that the three protection officers knew to expect a handover of papers, and that he had personally sanctioned the papers, clearing his actions with the Americans and enabling Mitya to pass them up the Russian chain of command.

Within three weeks, there were indications that the Soviets' focus had moved abruptly away from weapons development. The Space Race began in October 1957, when the Russians launched their R-7 rocket from Baikonur, with a satellite, Sputnik 1, attached in the nosecone – the place a warhead might otherwise have been. The international community were surprised, impressed, and relieved at the direction the rocket had taken.

Two months later, the Americans launched their own satellite, but their attempt failed and ended in an explosion just 4 feet up from the launch pad. In February 1958, they tried again with Explorer 1, which returned data from space for four months.

Civilian Life, 1960

In 1959, Eric left Germany, moving first to Britain and then to America in 1967. He moved with Marianna and their children to work at the British Embassy in Washington, Maryland, USA. He had been offered a job with the Skynet Defence Communications Satellite Project, and this was a chance for him to bring up his family in the 'land of opportunity'.

Before Eric left Gatow, there were several conversations with Cocky, designed to help him adjust to civilian life in the UK. My father moved back to the UK in 1959, finally laying down his false identity, and exhausted from years of tension.

Eric had impressed on Cocky the necessity to marry someone, start a family, just as he himself had done. He had a word with his contacts at the Air Ministry and managed to secure a job interview for Cocky at Pye Electronics in Cambridge.

'You're going to need a job, a house, a car and a wife,' he told Cocky. 'In that order.'

Cocky had laughed. It all sounded so simple, and for Eric, it really did appear to have been straightforward. Throughout all his exploits, Eric had Marianna in the background, but Cocky was starting from a position of entrenched singleness.

* * *

In 1960, Cocky shed his RAF nickname and reverted to his real name of Neville. He started work at Pye Electronics. Working again in an ordinary company proved difficult because the expertise he needed was at a low level compared to his experience. He couldn't tell the truth on his CV, nor in any of the work-related conversations he had every day.

The job came with a house, and his line manager, uneasy at his single status, told him that a company house, with four bedrooms, would be available for him if he were to marry. The prospect of marriage, though, seemed like a faraway country he could never reach. He knew no one in the UK other than his mother, and his brother and sisters.

One Saturday morning, he made his way to his older sister Marjorie's home in Bristol. Marjorie gave Neville a huge hug on the doorstep, her children around her and her husband Bill out doing his rounds. Just having his warm, affectionate sister in close proximity, let alone receiving an actual hug, was a strange feeling for Neville. He didn't know what to do with his arms, where to put his feet, and now, with Marjorie's bosomy body up against him, the sister that he knew so well from years of shared family traumas, he was willing the hug to be over.

'How's Mother?' he asked as he gently pushed her away. 'Is she still in her own place?'

'Well, I must tell you,' said Marjorie, 'she does still have her own flat, but she has gone back to Iggy.'

'That's disappointing,' said Neville. 'I was hoping she would find a better life without the old man.'

Neville and Marjorie pushed through to the kitchen, while the children danced and scampered around them, shrieking and trying to carry Neville's shiny new briefcase.

'Mother and Father turned up on my doorstep a few months ago,' said Marjorie over the top of the children's noise. 'Bill was here, and he said Mother was welcome to come in, but not Iggy. We talked on the doorstep for a while, but Bill absolutely refused to let Iggy in, so they ended up going away. When I spoke to Mother about it, she said that her marriage vows held her to stay with him, and she told me his behaviour had changed for the better since she went back.'

Neville looked unconvinced. So many times his father Iggy had sworn he would change, but the drinking and the violence would start up again at the smallest provocation.

'She's keeping her flat on,' continued Marjorie. 'It's her bolt-hole; Iggy has no idea where it is, even though he keeps on trying to get her to tell him the address.'

'That's something, then. I only hope she can keep that secret,' said Neville. 'It gives her an alternative place to go, somewhere away from him.'

'She really thinks she is the one at fault,' said Marjorie. 'She believes that marriage vows are totally binding, and that she is the one who broke them by moving away. She can't accept that Iggy broke those vows again and again.'

'I hate him,' said Neville simply, 'for what he put Mother through.'

'And us,' said Marjorie. 'Don't forget us. We children didn't have any sort of childhood, did we?'

'No, I can't remember any happy times,' agreed Neville. The misery he felt threatened to overwhelm him. He changed the subject. 'But you and Bill are doing well?' He looked at the children still careering around the kitchen. He wouldn't know where to start with talking to them. They were alien and unruly.

'I hope you will meet someone to marry', Marjorie was saying, 'who can show you how to do this family life thing. Believe me, it feels great when we are all around the tea table, talking and eating and laughing together.'

'I can hardly imagine being married,' said Neville, 'let alone having children. I've been this other person, a sort of walking brain-box, called Cocky, for so long now, I've forgotten who I really am.'

'Your RAF nickname is Cocky?' said Marjorie. 'That's hilarious. Cocky, Cocky Cocky – no, I don't think it suits you.'

'I've ditched it. I've come back to this country now. I'm finished with the RAF, and all I want to do is start some kind of settled life.'

'Getting married fits with all that perfectly,' said Marjorie.

'Yes, but I don't know how,' he replied. 'And I don't think that anyone would be in the least bit interested in me. I'm 36, for a start.'

'It's alright for men to be a little older,' said Marjorie over their cups of tea on the kitchen table. 'It's the women who have to watch their biological clock.'

'How come you know so much about it?'

'I met Bill in the nick of time. I was 26, he was 32, and we had the three children straight away.'

The children were sliding on the stone-flagged floors and whipping each other with tea towels. Marjorie gazed benevolently at them. She knew that when Bill came in, he would sweep them up in his arms one by one, and then set them down gently so they could continue playing.

It was hard for Neville, though, to get the time and attention he needed from his sister.

'My friend Penelope', said Marjorie, 'met her husband through a Marriage Bureau[2] in London. She found a most suitable person to marry, in the minimum of time.'

'That sounds gruesome,' said Neville. 'Like a human lottery.'

'No, it's not at all gruesome. It's practical.'

'The man, and the woman, are interviewed separately by the people who run the Bureau. They ask a lot of questions and write everything down. They ask what you are looking for in a marriage partner, what age, what type of person, anything you particularly like or don't like. There are lots of people signing up to be on the Bureau's books, and then the person who interviewed you writes to tell you that a particular young lady is keen to meet you, and you get their telephone number, or their address to write to. You don't have to marry the first one you meet. But at least you will know she is properly available.'

'Just knowing that someone is free is really hard,' said Neville. 'I never seem to get close enough to see if someone has a wedding ring on or not. And many women nowadays have a beau, but they don't walk out together.'

'Exactly,' said Marjorie. 'With this Bureau, you would only be introduced to someone who had stated that she is free to marry and is actively looking for a man just like you.'

Neville couldn't imagine anyone stating that they would like a man just like him.

'And you know that the people at the Bureau are experienced in these things, so the chances are much higher that you will like each other.'

'It could work for me,' said Neville wearily. 'I don't know how to go about meeting women. Seriously, I haven't talked to one properly since ... Singapore.'

'You've got a lot to offer a woman, Neville.'

'I'm not sure,' mused Neville.

'You're better dressed than you used to be. And, you've got a car, haven't you? Not many men can afford one of those.'

'I'm not sure, though. Women today are so ...' mumbled Neville.

'It's 1960,' declared Marjorie, 'and women today are modern. They want a husband and they want a job of their own too. When our children are at school, I'm going back to being a typist. Women today want a husband who will help with the children, put them to bed and all of that. They want a man to talk to as well, not just keep house for.'

Neville doubted his ability to find enough to talk about with a woman. He had no concept of what putting a child to bed might mean, when all he knew was that children took themselves off to bed sharpish when Mother shouted 'bedtime', minutes before Father came in from the pub.

'I've started at Pye's in Cambridge. It's a good job, I suppose. That's important, surely?'

'Yes, a job, of course that's important. But, there is something that Bill does – he comes in at the end of his working day and he has this, well, it's a sort of blast of happiness as he walks in through the door. He can't wait to see us all, and he can't hide it, he's just ... warm and ... wonderful.'

Neville had never felt less warm or more removed from being wonderful. He couldn't recognise any future version of himself in the loving man that Marjorie was describing. This marriage thing was a complex challenge that he didn't have the skills for. It had unwritten rules. It was vague and unscientific. If only marriage was as straightforward as rocket science.

'I did meet a woman once ... who was someone who ...' he started.

'Go on,' said Marjorie.

Neville had always to curate his memories. He could not mention the whole swathe of time he had spent in Obernkirchen, Gorodomlya, or Gatow – these areas of his life were subject to the Official Secrets Act. If he was going to mention Kaarina, he would have to go into the specific strangeness of the way they had met, and the reasons why, even if she had returned his feelings, which he doubted, they could not now be together. It was hard to be honest with the people he cared most about, such as his sister, when he couldn't refer to the events of a decade of his life. He clammed up.

Marjorie handed a crust of bread to her child who was teething, and thought about her brother. She could tell he didn't have his usual confidence. He was brittle, wary, and closed in. After spending half an hour with him, she agreed he would struggle to meet someone at the moment. She pushed an envelope towards him. A name, address and telephone number were carefully written on the back.

'The Marriage Bureau,' he read aloud. '124 New Bond Street, London.'

Marjorie was picking up a child who had crashed into a coal bucket. She pulled him onto her lap and was rocking him, while trying to continue her conversation with Neville.

The children's noise suddenly seemed unbearable to Neville. He finished his tea and put the mug down on the table. 'Thanks, Sis.'

He put the envelope in his pocket, and kissed his sister goodbye. He set off for his hotel room in Bristol. He was looking forward to a little peace and quiet. The first thing he would do, he decided, would be to get a better haircut. He would give the job at Pye's his best shot, and have another look at the address in his pocket and think about whether to pick up the phone.

Three weeks later, Neville made his way to New Bond Street. He was smartly dressed and had that morning bought himself another car – the Alvis

– from his brother. All he needed now to complete the circuit of things that when wired together formed a civilian life, was a wife.

Whenever he thought about the word 'wife', the strongest image that came to his mind was always of Marianna, serving goulash to him and Eric, her belly swollen with their first child and her devotion to Eric palpable. The next image was often Irmgard, her smile full and warm, like some kind of angel, and her delight at giving him delicious German meals a balm to his spirit. The third and most abiding image to come to mind was always Kaarina. Her blonde hair and green eyes, playful smile and sweet seriousness did not automatically spell 'wife' as much as 'soulmate'. She was exotic and unreachable. He had no idea how in love with her he had been through those frozen months on Gorodomlya. He repressed it, and tried to forget her.

He pushed Kaarina's image away now as he used the card Marjorie had given him to find the correct street address. Number 124. He climbed the stairs of the tall Bond Street building, and was immediately mesmerised by a riot of swirling colours and the excited chatter of four young ladies emerging from a door on the second floor. Wearing outrageous hats, they were shrieking with delight at the sight of each other in their new millinery. One of them pulled the brim of her hat down when she saw Neville, and winked suggestively at him with her film-star lashes. He realised they were emerging from a hat-making business on the floor below the Marriage Bureau.

'Don't,' said one of the ladies, wagging her finger at him. 'Don't go in there. Do not sign on the line, or, mark my words, your life will be *over*. Marry me instead.' She hooked her arm into his and peered up at him, just as she had seen ladies do in films.

'Or marry *me*,' said another, taking his arm possessively.

'Ooh, don't listen to Evie, she's already married!' shrieked another.

'I'm free,' purred a dark-haired woman in a cloche hat that went superbly with her smooth, dark bob. She took his other arm so that he was, for a moment, dripping with adoring women.

Upstairs, the Marriage Bureau door opened just a crack, and an older lady with grey hair and a friendly face called back into the office: 'Heather? Mr Cox, your two o'clock, is here!'

The ladies giggled and latched on even more tightly to Neville's arms.

'Oh, won't you be *my* two o'clock?' said the first lady, who had dark bouncy curls set against a pink outfit and pink lips drawn in a bow.

'Come to the races with us on Saturday afternoon,' purred the smooth-haired one in the cloche. She had a low voice that reminded Neville of Kaarina.

'Epsom Races,' shrieked another. 'It's only an hour's drive from London.'

'Do you have a car?' asked the one in pink. 'You look like you might have a car.'

'I'd love to come to the races with you,' said Neville truthfully. 'I really would.'

'Do come up,' came the friendly voice again from the floor above.

Neville took a step towards the stairs. This loosened the ladies from his arms, and they let him go with mock reluctance and clattered down the stairwell, holding their new hats on their heads and still shrieking.

'See you on Saturday!'

'Bye,' he shouted after them.

Neville climbed the stairs, his head reeling.

'I'm putting my money on the two o'clock,' came echoing up the stairs, accompanied by gales of distant laughter, before the street door closed and the stairway went quiet.

The inside of the Marriage Bureau office seemed sober and businesslike, especially so after the colour and fun that the ladies had created for those few memorable minutes. The friendly older lady had disappeared into the back room and Neville was faced with a woman who was taller than him, statuesque, and possessed of an air of confidence that made him feel inadequate. Heather Jenner was noticeably elegant, beautifully dressed and made up, with an upper-class voice. She introduced herself as Miss Jenner, shaking his hand and indicating to him where to sit, all in one practised gesture.

As he settled into his seat, conscious of sitting up straight, as he was sure she was about to remind him, one part of his brain was calculating the logistics of getting to Epsom Races on Saturday. The other part was feeling viscerally lonely.

'We quite enjoy having the milliners on the floor below,' Miss Jenner said to him conversationally. 'It's endlessly entertaining.'

'It must be,' agreed Neville.

Miss Jenner looked through some notes that her assistant had taken down over the telephone.

'You're just out of the RAF.'

'Yes,' said Neville.

'You're 36.'

'Yes.'

'You've got a job in Cambridge, a house, and a car,' she read.

'Yes, an Alvis,' said Neville, 'with leather upholstery.'

'What do I call you? Is it Neville, or Mr Cox?'

Neville had the distinct sensation that all this was happening to someone else. He was tongue-tied and had no idea what the social rules might govern the correct form of his name.

'I'll stick with Mr Cox,' said Miss Jenner briskly.

'Yes, please do. I had a nickname for a long time. But not anymore.'

'You will need to fill in a registration form. We'll do that in a minute. But for now, please tell me in your own words about yourself, and about the person who you are looking to marry.'

'I've been in the RAF since 1946, mainly in Germany, but I was also in Singapore for four years. I'm a draftsman by profession. I'm back in the UK for good. I've got an engineering job at Pye's in Cambridge, which comes with a house. I've got a car, as I said, and now I'm looking for a wife.'

'You've told me all the facts that I already have written here,' said Miss Jenner, 'but I still don't know anything about you as a person – your expectations of a wife, or what you would bring to a marriage, your hopes for the future.'

'My boss at Pye's thinks that I'm more likely to get a promotion if I am married. And I would get a bigger house – a family-sized one.'

'A family-sized house. Are you wanting to have some children?' probed Miss Jenner, slightly impatiently.

'I just want to have an ordinary life like everyone else. Do a good job, drive my Alvis around, have a family too, a wife, someone to come home to at the end of each day.'

'Someone to come home to,' echoed Miss Jenner. 'Tell me more about that. What is she like?'

'She's younger than me,' faltered Neville.

'Uh-huh,' encouraged Miss Jenner.

'She's, um, a good cook,' said Neville. 'There's a good meal waiting for me when I come home.'

'You're looking for a domesticated sort, then,' clarified Miss Jenner. 'I'm not sure if …' Then she decided to pick up on something even more concerning than Neville's assumption about the good meal waiting for him when he came home from work.

'Does your boss, um, is he insisting you get married before he will offer you this promotion?' she asked.

'It's implied,' said Neville.

'Do you always do what your boss wants you to?'

Neville thought for a while about this question. 'My boss in the RAF suggested I come back to the UK. And my new boss at Pye's mentioned that my package would include a bigger house if I had a wife and family. All I want is to disappear. Into normal civilian life. Not stand out in any way. Just be a UK citizen with a house, job, car, and …'

'Wife,' said Miss Jenner. 'Quite.'

There was an awkward silence. Neville wondered if the depth of his despair was visible to this very self-possessed person in front of him.

'Mr Cox, I'm not sure we can help you. You see, the young ladies who sign up with this bureau want, even *deserve*, so much more than you appear to be able to offer. They do not want to be your wife so that they can provide you with meals. They do not want to complete the set of components of your 'civilian' life. I'm sorry, Mr Cox, I cannot think of a single person we have on our books who I could in all conscience put you forward to meet. Many young ladies want security, that's true, and your good job is a mark in your favour. But you are not thinking about them at all, why they would choose you, what sort of a life you could build together. I'm getting a bit of a bleak feeling from you. I wonder if your heart is in it. No one has ever told me they want to marry because their boss thinks it a good idea.'

Neville sat silently, looking at his hands.

'Mr Cox, do you really want to get married?'

'I don't know,' he answered miserably.

There was a silence, during which Miss Jenner was trying to find the words to end the interview.

'I would never raise a hand against a woman,' Neville volunteered, hoping to tell her a little of who he was, but this statement made things worse.

'Of course you wouldn't; the very thought!' said Miss Jenner sharply.

'Would you like a cup of tea?' asked the grey-haired assistant from the back room. Neither Neville nor Miss Jenner answered. A cup of tea would prolong the interview, which was coming to a natural end.

As so often happened when Neville was feeling down, an indistinct image of Kaarina's green eyes and blonde hair came wafting towards him. She was smiling warmly, waiting to shake his hand as she had done after the concert on Gorodomlya. Her face seemed to promise the world, starting with the first genuine female friendship he had known in years.

He was moved to speak.

'I want someone, a good friend. I do want someone to share my life, and me to share her life,' he began. He imagined Kaarina playing her cello, maybe getting ready for a concert, while he pottered around in the kitchen, possibly getting a meal for the two of them, or even feeding a couple of young children, children with bright blonde hair like hers, giving them their tea so that Kaarina could get ready. She would need to look smart for the performance, and he would be driving her there, proudly dropping her off outside the concert hall in the Alvis. She would kiss him goodbye before heading up the steps, holding her cello in its case.

'What would sharing someone else's life look like?' asked Miss Jenner.

'Being interested in her, her interests, helping her to succeed, listening when she's had a disappointment, driving her to places. If she wanted children, then yes, having a couple of children, and I would look forward to coming home to her, and if we had any difficulties then we would sit down after the evening meal and not put the TV on or listen to the radio. We would sit down together and talk and make things better, work out a plan ... we would talk, all the time.'

Miss Jenner could hardly believe the change in Neville's ability to express himself. 'Oh, yes please, a cup of tea would be lovely, thanks, Dorothy,' she called into the back room. 'And one for Mr Cox. We're just going to fill in his registration form.'

The image of Kaarina and her cello disappeared from his mind, and the next half hour was spent filling in the Bureau's form, listing his parents' occupations as Merchant Seaman and Housewife.

'Merchant Seaman,' read Heather, as she mentally categorised Neville's background as working class. 'That's a responsible job. You must be proud of your father.'

'I'm much more proud of my mother,' he replied.

'"Housewife" – well, yes, always a challenge, keeping everything spic and span,' said Miss Jenner briskly.

Neville's eyes filled with tears at the thought of his mother's determination to bring up six children against all the odds. 'Housewife' came nowhere near expressing the sheer doggedness she had needed to get the family through each unpredictable day. He would never put his own wife through such a treadmill of a life. He would do everything in his power to be the husband that she needed. He would provide food and a house that she could make comfortable – and even children if she wanted them. He would do that for her.

Miss Jenner was still talking. 'You seem very concerned with meals, Neville. Would you expect your wife to have a job of her own? Or do you see her as a homemaker, looking after the home and children and preparing your evening meals? Would you ever be the one to prepare a meal – for example, if your wife also had a job?'

'I can't cook. I would expect my wife to do that. I've never had to learn.'

Miss Jenner pursed her lips at his flat refusal to consider sharing responsibility in the domestic sphere. He had sounded quite modern when he was talking about supporting the wife he wanted, but clearly, he was stuck in a traditional mindset.

'And what amount do you earn? asked Miss Jenner, for the form.

'With this next promotion, it'll be £2,500 a year,' said Neville.

Miss Jenner looked disbelieving. The person in front of her, despite his new clothes and shiny briefcase, didn't look capable of commanding that sort of salary. He might be one of the men on the books who exaggerated their assets. Their would-be wives sometimes wrote her aggrieved letters, saying such things as:

His 'holiday home in France' is a caravan on the south coast of England. You can just about see the coast of France on a clear day when there's not too much fog.

His five-figure salary starts with three zeros ...

The open-topped car certainly has an open top. It is in fact a bicycle, with one of those sidecars attached.

Miss Jenner tapped her pencil for a while, until she had to write Neville's salary down as he wasn't amending or retracting the figure.

'I was a sergeant in the RAF, and I kept up with advances in engineering, which is coming in useful now I'm back in Civvy Street,' he said.

Miss Jenner had to accept his explanation.

'And do you have interests other than cars?' she asked him.

'Classical music,' said Neville.

'Religion?' asked Miss Jenner.

'C of E,' said Neville.

The form now complete, the older lady, who introduced herself as Miss Harbottle, came in and stood holding the teacups. Miss Jenner stood up to indicate the interview was over, shaking Neville's hand again.

'We'll be in touch as soon as we can,' she said, 'by telephone or letter, to let you know of any matches that we find.'

'We've got several suitable ladies on our books,' said Miss Harbottle. 'We hope there is someone there for you.'

Neville smiled at them. 'I hope so, too,' he said, as he left the room.

'Any good?' asked Dorothy Harbottle when his footsteps had faded.

'Doesn't really know what he wants,' said Heather. 'He's horribly focused on meals. I mean, many people who lived through the rationing of the war years care about food. That's normal. But for Mr Cox, it just seems to be too important. He expects a wife, a modern wife, probably with a job of her own, to be waiting for him each evening with a home-cooked meal!'

'Women having a job of their own is something that the war years brought,' said Dorothy, 'but I don't think that men have really cottoned on to what that means for family life. It's still normal, now in 1960, for wives to do the cooking, cleaning and childcare, even if they do go out to work.'

'I couldn't do all that domestic work without help,' said Heather. 'It sounds exhausting.'

'Plenty of women do it,' said Dorothy.

'I can't think of anyone, off the top of my head, for Mr Cox,' said Heather. 'Usually by now I've started to form a few ideas, but this one's difficult.'

'He seems unhappy.'

'Humph,' said Heather. 'He seemed guarded. Unsure of saying the wrong thing. And what was that about not raising his hand to a woman in anger?'

'I don't know. Some men do have a temper, and there are even those who hit their wives. His father was a sailor, remember?'

'Hit their wives? God forbid!' said Heather.

'He seemed rather attractive to me, when I saw him with those ladies on the landing,' said Dorothy. 'He's nice and slim, and he's got lively, intelligent eyes.'

'Has he? He seems on the lost and miserable side to me,' said Heather, before taking Neville's registration form into the back office. 'Look at this,' she exclaimed. 'In the "Wife wanted" section he has put "Younger, good cook, *goulash*, good conversationalist". What on Earth does "goulash" mean?'

'It's a Hungarian dish, a sort of tomato-y, peppery stew.'

'I know what it is,' snapped Heather, 'but what does it mean here? He wants someone to cook him goulash? He can't make it himself, so that's got to be it. This beggars belief.'

Dorothy and Heather giggled. They both knew that from this point on, 'Goulash' would be their shorthand for Neville, differentiating him from the other ex-services men on their books.

'Goulash,' said Dorothy to herself. 'Let's have a look through our ladies' card index box for someone for you. I think there are still some happy homemakers out there.'

* * *

Once outside, Neville glanced up and down the street, hoping that the millinery ladies had stopped for tea and cakes at the Lyons' Corner House. If they were there, surrounded by brightly coloured cakes and still shrieking with laughter, he would walk in, and arrange properly to drive them all to Epsom Races on Saturday. There was no sign of them when he peered in through the window. A waitress inside looked uncomfortable at him staring in so intently.

When he got back to his brother's house in London, he picked up a newspaper and sat at the kitchen table with his head in his hands, not reading it. He couldn't find the carefree young man he had been. He tried to reach the person who had gone to Singapore and sailed, and played games on the beach with the locals. There had been some definite chances at romance, and it had all seemed so easy.

The RAF wasn't just overwhelmingly male, it was also an environment where you had to keep all your feelings and the majority of your thoughts to yourself. Even those friendships at the Unit in Germany had been fleeting, with Marianna the only female he had known. On Gorodomlya, Irmgard and Kaarina had, in their different ways, made life worth living. But Irmgard was married and Kaarina, as far as he knew, had never seen him as a marriage prospect.

Sitting at his brother's table, Neville reflected on how the Cold War had distorted his life. It had been impossible to follow his own path. Impossible to be an engineer, a scientist, or a young man in search of a wife. You were there to avert danger, not get yourself killed, and do the little that you could to prevent the world imploding.

The time on Gorodomlya with the Finnish quartet had been an oasis of normality – in the company of attractive young people, talking about music, laughing, and joking around in a way that was a revelation to him. For the first time, he had been someone warm and interesting, even someone who had inspired his friends to help him make that difficult escape. His eyes stung as he remembered how Kaarina had pretended to faint, just at the critical moment when his inability to play the violin at the Vaalimaa border was about to become evident. She had saved his life, putting herself at risk for him, and he would never now be able to tell her how grateful he was. Just briefly, on the landing with the milliners, he had experienced a little of that spontaneity again, had glimpsed himself as someone worth knowing – even to highly attractive young ladies.

* * *

He thought about his brother Bernard, his younger brother, who had died aged 4. Marjorie had mentioned recently that now she was an adult, she was ever more convinced that Iggy had been responsible for Bernard's death. Neville remembered the day in question. It was etched on his memory, then buried in a 'do not open' mental filing cabinet. He opened it now, gingerly, ready to slam the lid down. His mother had wanted to take all six children around to see her own mother, Nanny Becky. They planned to walk through the park. Iggy was in the house, in a bad mood. The main reason for the outing was to keep everyone safe from him. The children had learned to fear his unpredictable rages and a walk to Nanny Becky's was preferable to staying at home.

Bernard had been whining for his scooter ever since the outing had been mentioned. A neighbour had given him a scooter the day before, and he wanted to ride it through the streets. As the family left the house, he wailed that he needed someone to open the shed door, down a flight of steep stone steps. Neville wanted to go with his brother to open the shed door, but his mother insisted that Bernard could get the scooter on his own. They had watched the little boy walk back down the street, and watched and waited for him to reappear with the scooter. He didn't reappear, so Neville set off, thinking to help Bernard with the shed door. When he reached the stone steps, he saw Bernard at the bottom, crumpled in a heap with his head cut open.

Iggy was in the house. But Neville called for his mother. He went back to the family group waiting on the pavement, crying that Bernard had fallen down the steps.

Bernard never recovered; he had died on the spot.

Iggy never left the house through all of the commotion – the ambulance coming, the neighbours flocking round, Neville's sisters weeping and trying to get a response from the little boy. Bernard hadn't reached the shed door. It was still closed, and he was facing forwards.

When the family finally went back in the house, Iggy had gone to the pub, and they didn't see him for days after that. The coroner recorded an accidental death. The family went on with their chaotic lives. But Neville never forgot his brother's knees, skinned and dirty as they always were, a 4-year-old's knees.

Even now, when he pulled on long trousers, his dead brother's dirty knees would come to mind.

Marjorie's theory was that Iggy had heard someone coming back to the house and had assumed it must be Mother. He had come out wanting to hurt her. When Iggy saw it was Bernard, he was unable to stop his angry impulse and had shoved him, his adult strength far too much for the small boy. Whether Iggy had meant to kill Bernard, Marjorie doubted, but she had Iggy down as the cause of his death.

For Neville, one of the worst aspects of the aftermath of the tragedy was that hardly anyone had noticed Bernard's disappearance. None of the teachers at school, the neighbours or even their own wider family had an idea of who Bernard had actually been. He was just one of six children, one of three boys, and his disappearance caused hardly a ripple.

When Walter had gone missing from the laboratory on Gorodomlya, people wouldn't admit openly that his 'disappearance' was an orchestrated effort by the Russians. It had been chilling to hear 'No one by the name of Walter works here, or has ever worked here' when he knew full well that Walter was a brilliant, funny and humane person. He made an effort to remember Walter every day, just to prove the Russians wrong. Remembering Bernard was different. He would wander into Neville's mind at almost any time.

His brother John came into the kitchen, breaking Neville's downward spiral.

'Hello, Bro,' he said lightly.

'Hello, John.'

'We are thinking of having a day out with Mother and the girls tomorrow,' said John. 'Thought we'd make a day of it at the beach, take a picnic and some games. Want to come? We'll need both cars.'

'I was thinking of going to the races at Epsom,' said Neville.

'Going to the races at Epsom? Whatever for?'

'Just a whim. I fancy doing something fun.'

John sat down opposite Neville and looked into his strained face with affection. 'What do you say, to me and you going to Epsom races together another time?' he suggested kindly, concerned that Neville really shouldn't be on his own.

'Alright,' conceded Neville reluctantly. 'A day out with the family sounds good to me.'

'Good man,' said John, squeezing Neville's shoulder, and he went to tell Lilian that Neville was up for the family day out.

Neville heard Lilian's delighted response and her immediate switch into organising mode: picnic, tennis racquets, buckets and spades, and blankets for their mother – everything the family needed to be away for a few hours. He sank back on the kitchen chair, his head in his hands.

Civilian Wife

In the house in Cambridge, Neville had bought a few pieces of furniture and was trying to see the place as home. His new job consumed only office hours, something he wasn't used to at all. Coming home in the Alvis and opening the door to empty rooms, his breakfast things exactly where he had left them, was depressing. He had no idea how to shop for food, as he had never had to do it.

Before long, he had taken on a cleaning lady, Mrs Perkins, who arrived in the mornings as he left for work. She had started writing him shopping lists of food items to buy. Then, almost imperceptibly, she had started doing his shopping. It wasn't long before she was also leaving a home-cooked meal waiting in his newly cleaned kitchen, something tasty that just needed to be warmed through.

Occasionally, the couple of hours she worked for him were earlier in the day, before he left for work, and on one of the occasions when they coincided, Mrs Perkins pointed out a letter she had picked up almost a week ago from the brand-new doormat. She had placed it on the mantelpiece, but Cocky hadn't seen it. He didn't use his sitting room, because his music system was set up in the study.

'This came for you,' she said, holding out a cream-coloured envelope of thick, good-quality writing paper. 'I put it by your clock, but I don't think you've noticed it.'

The letter was beautifully addressed to him in handwriting he didn't recognise. Mrs Perkins was hovering, waiting to see if he would tell her anything about the letter. She loved a bit of intrigue, and getting insights into the lives of her clients was one of the few perks of her job.

Cocky thanked Mrs Perkins, slid the envelope into his inside pocket, and picked up his briefcase. He walked calmly to the Alvis parked by the kerb outside. Before starting the engine, he pulled the letter from his pocket and ripped it open. He read it through several times, his heart pounding with excitement.

Dear Mr Cox,

Many thanks for visiting the Marriage Bureau recently. We enjoyed meeting you.

You will be pleased to hear that we have someone recently on our books who we believe you would like to meet. She is 26 years old, a teacher, 5' 10" tall, and she has brown curly hair and blue eyes. She meets most of the stipulations you mentioned at interview.

We enclose a slip with her name, address and telephone number. She has given permission for you to contact her at your convenience.

Yours sincerely,
Miss Heather Jenner

As Neville drove off in the Alvis, feeling more hopeful than he had done in a long time, he failed to notice a big black car parked on the other side of the residential street, a few doors down. The occupant of the car was wearing dark glasses and scribbling in a notebook, and as Neville drove away, he reached for his camera on the passenger seat.

Mrs Perkins came out of the front door with a metal pail half-full of vinegar and water, and started to clean the windows with a chamois leather she had found in the garage. The stranger in the car took plenty of photographs of her as she stretched to reach the top of the windows and bent down to wring out the cloth, and finally, one of her face when she glanced around briefly before going back inside.

By the following week, Neville had arranged a meeting with the young woman mentioned in the letter from the Bureau. It was to be on the Saturday coming. The plan was that he would drive over to her flat in Cookham in the

Alvis, and take her out for a picnic in the countryside. Each workday for the rest of the week felt like an eternity. Every trivial question at work annoyed him, and even though he tried to work late just to shorten the lonely evening hours, there just weren't enough tasks to keep him absorbed. He had never known days to drag like this, and spent the evenings making sure the Alvis was gleaming – even polishing the leather seats.

Jean was waiting for him outside the house where she rented the top-floor flat. She was tall and smiling and her curly hair blew around in the wind. She was trying to hold it off her face while gripping a small picnic basket and a striped rug. Neville parked next to her, right outside the house.

'Hello. You must be Jean,' he said. 'Brown curly hair, and 5 feet 10 inches.'

'And you must be Neville,' she laughed. 'You've driving an Alvis.'

'She's my pride and joy. I've polished her up for today.'

'Very nice,' said Jean, handing him the picnic basket and rug.

He placed them in the small boot and turned around to see her disappearing back inside the house.

'I'm going back inside for a headscarf,' she called. 'Won't be a minute.'

She was much longer than a minute, and Neville started to wonder if she had changed her mind. But he had possession of her picnic basket and rug, so he waited outside. As he waited, he resolved that he wanted to get to know her; he wanted the day out to proceed as planned.

Inside the flat, Jean was less sure, and apart from finding a headscarf, she was wrestling with an attack of indecision, tinged with disappointment. She wasn't sure why, but Neville wasn't the man that she was expecting. She had to tell herself that this was one day, one drive, one picnic – it did not mean forever. Afterwards, she could come back here and recover. She glanced out of the window. He was waiting, so she tied on the headscarf and went back outside.

Neville opened the passenger door for her and she slid across the seat. It really was very slippery. 'Thanks for waiting. It took a while to find the right one', she said, indicating the headscarf, 'to go with my coat.'

'Well, it does go nicely with your coat,' said Neville, relieved that she was finally sitting next to him.

'Did you find my house alright?' asked Jean. 'The road signs aren't really very good. Lots of people drive straight past Cookham and end up in Maidenhead.'

'I've got a road map,' said Neville. 'If you look in the glove locker, which is right by your knees, the Alvis has a really good, big locker space to keep things in, so the car doesn't get cluttered up.'

Jean duly peeped into the glove locker, but could think of nothing to say about it.

'How long have you lived in Cookham?' asked Neville, who was keen to fill the silence. He had practised some questions on the drive over, just in case he needed to have a few ready.

'Three and a half years. Since I qualified as a teacher,' said Jean. 'It's convenient because I live right next to the school where I work. It's a good thing on one hand, because I don't need a car, but it has its downside, because all the children know where I live, and I see them with their parents at the local shop, and at the church. They all call me "Miss" and it seems to entertain them to see their teacher doing ordinary things – like going for a walk or hanging out washing.'

Neville wasn't sure why this was a 'downside' for Jean. It seemed to him quite charming that the families of the village had their teacher living amongst them. But he didn't ask about it, preferring to move to his next prepared question.

'Where do your parents live, Jean?'

'In Slough … you know, it's in Berkshire. Some people say "Windsor" rather than "Slough", but I like to say it as it is.'

'What does your father do?'

'He fought in Burma during the war; he had a terrible time there, although he's never actually been able to talk about it.'

Neville said nothing. Jean was compelled to continue.

'My father, George, has gone back to work at Mars chocolate factory in Slough. You would think that he could easily pick up the threads of his old life, but he's never really been the same since he got back from the war.'

'And what about your mother?'

'What about yours?' asked Jean, with a flash of impatience, as she tried to stop the flow of questions. This felt more like an interview than a conversation, and she really didn't want to disclose things about her mother to this person she had only just met. She noticed his hands. They were small and neat and clothed in smart driving gloves. They gripped the steering wheel evenly, assuredly, and made quick, neat movements every time the car turned right or left.

'My mother?' asked Neville.

'Yes – I'd like to know about *your* mother,' said Jean sulkily.

'She lives in Bristol,' said Neville. 'Where I come from.'

'You're not from Cambridge?'

'No. I just happen to be living in Cambridge. Where my job is at the moment. I was born in Bristol. My mother and father still live there. Some of my brothers and sisters still live there too.'

'How many brothers and sisters do you have?' asked Jean. This was tortuous. Like a beginner's French language class, except they were supposedly speaking the same language. And her watch showed the time was still only two o'clock.

'I'm one of six children,' said Neville.

'Oh,' said Jean. She knew it would be easier, friendlier, to ask about the brothers and sisters, and it would fill the silence. But Neville seemed only capable of telling her bare facts, and he hadn't picked up on her conversational offers. He seemed to be closed up, secretive even, only giving her the minimum of information as if his life depended on staying unknown.

Jean had already decided to get the dull picnic over with, get back to her flat, and contact the Marriage Bureau to ask them to find her another match. She sat in the passenger seat, thinking about the person she was hoping for. Someone taller than herself, someone warm, someone who thought he had struck lucky to be allowed to take her out. Someone able to take a risk, give a bit of himself away, help her feel in some way important to them. The very least, she mused, was someone willing and able to find out who she was, beyond being a teacher with a mother and a father.

Hot tears were forming and threatening to escape down her cheeks. She wiped them with her scarf and took a surreptitious look at Neville's face in profile as he concentrated on driving the car. He wasn't at all bad looking. She could see that they might be evenly matched in that respect. But he was so self-contained – cold, even. He certainly didn't seem to need anyone else.

Just one drive, one picnic, she thought to herself grimly.

'We're here,' said Neville after a while, his voice a little unsteady because he could tell that the day out wasn't going very well so far, and he wasn't sure why. He wished he didn't care, but to his surprise, he found himself caring very much, and wanting to find the key to changing this stilted date into a success. Jean was very attractive – stunning, even. Perhaps she was out of his league. He didn't know. He only knew that he couldn't think of a thing to say that would change the day for the better.

'It's a bit cold for a picnic after all,' said Jean, thinking that actually, eating the picnic in her Cookham flat, on her own, might be preferable to prolonging the time with Neville.

'There's a bandstand – we could find the sheltered side of it and spread out the rug there,' said Neville. He wished he had suggested a theatre trip, or a concert, some warm place where there wasn't so much pressure to make conversation.

Jean laid out the rug beneath a wooden slatted seat and set out the picnic food. She had made sandwiches and there were boiled eggs and tomatoes, and a packet of biscuits. She had bought a flask for tea from the local shop, and she thought now that buying the flask had been a waste of her meagre teacher's pay. She didn't want to sit on the blanket; it was enough to have Neville sitting fairly close on the bench beside her.

Neville said nothing about the food she had prepared.

'What did you think of Heather Jenner?' she eventually asked him. The Marriage Bureau, at least, was something they shared – perhaps the only experience they had in common. Surely, they could talk about that.

Neville was grateful for a second chance from Jean. He didn't want to blow it this time.

'She is formidable,' he said. 'She made me feel like a naughty schoolboy. I was sure she was going to tell me to sit up straight ... so I did: I sat up straighter than I ever have before.'

Jean laughed, relieved that Neville was able to give more than one-word answers.

Encouraged by her laughter, Neville went on. 'She is like no one else I know. Obviously, she's upper class, and I'm most definitely not from her class, but she just seemed to have no understanding of what it might be like to have fought all your life to leave poverty behind. She asked about my mother, and I said *Housewife* – you know, the question about occupation.'

'I thought they only asked for the father's occupation,' said Jean. 'That's all they asked me. It didn't feel very good me having to tell them that he was a factory worker.'

'Father's occupation, yes, that's how they decide which *class* you belong to. I wanted to tell them that my mother was so much more than a *housewife*.'

'I think they've matched us because we are both from the same class. We're both from "working-class" families,' said Jean. 'My mother – I *didn't* tell them this – is a school dinner lady ... because when she was at school herself she always got top marks, but she wasn't allowed to take the school exams because of being a girl. She was always very angry about that. And my father, the one who fought in Burma, used to be a miner in the Durham coalfields. He's 6 feet 4, and he hated working in the mines. He really was too tall for that. When he was 22, old enough to decide that he didn't want to be a miner, he got on his bicycle and cycled down the A1 main road, all the way to Slough. He went to sleep under a bush, and only woke up when someone kicked him awake. The man who woke him up offered him a job at Mars Confectionery.'

Neville relaxed a little. He liked just listening to Jean. It took the onus away from him. He didn't have to come up with questions. 'I can't imagine that Heather Jenner would know any factory workers, miners, dinner ladies, or merchant seamen in the normal course of her life,' he said. 'I think she married a Scottish landowner. There was something about it in the newspaper.'

'A Scottish landowner? I wonder if she met him through her own Marriage Bureau,' laughed Jean.

'She's hardly someone who would struggle to find a husband,' observed Neville.

'I was almost scared of her,' said Jean. 'I found myself putting on a posh voice – I've never spoken like that before, but there I was, *torking laik theece*.'

Neville and Jean laughed about that together, and it felt really good.

'Heather Jenner ... it wasn't that she was unsympathetic,' began Neville. 'She was alright, and I suppose she was just doing her job, finding out about the person in front of her in order to match them up with someone suitable. She just had no idea what it's like to be born to a family that ... you don't really fit with. She saw the word 'Housewife' and assumed she knew what that meant for the mother of a family like mine.' Neville was finally allowing himself to acknowledge the hurt that he had silently absorbed during his interview at the Marriage Bureau.

'I came away feeling royally patronised,' offered Jean. 'Like I was being judged and found wanting. Heather's assistant, the older lady, was much kinder to me. And I did need a bit of kindness at that point. But even she made me feel small. She came out of the back room, touched my arm, and she said, "Good for you. Becoming a teacher is a good step up."'

'I liked the older lady better too,' said Neville. 'She reminded me of my mother.'

'A kind person,' agreed Jean. 'I wonder if she had a hand in matching the two of us?'

'I'm an engineer,' said Neville. 'Perhaps they think we both made a "step up" from our lowly origins in coal mines and factories.'

They laughed again.

Jean had moved a little closer to Neville, but he didn't dare join her on the picnic blanket. She still had to raise her voice to be heard over the wind.

'There's an artist in Cookham, well, he's died now; he died just before Christmas last year. He was a local eccentric and we used to have good chats, at events like lectures and film nights. I ended up sitting next to him more than once. He was really short, and he had these thick glasses. He pushed a black pram around ...'

'A pram?' queried Neville.

'Yes. It was full of painting materials that he never seemed to use. Once, he had this shabby pram next to him, blocking the aisle, and people had to squeeze past it. I was sitting on the other side of him. Some people said he was well known, his name was Stanley Spencer. Anyway, that evening at the lecture, I told him that I was the local teacher, and he was quite interested in whether I had to go to university to do that. I told him my parents hadn't been able to afford for me to complete my School Certificate, which would have qualified me for university, so instead I went to teacher training college.'

'Go on,' said Neville, who didn't know who Stanley Spencer was, but he was still enjoying the ease with which Jean was keeping the conversation going.

'Well, this Mr Spencer, he got really heated – almost angry. He started talking really loudly, about "victims of the terrible English class structure". How it is full of "wealthy, titled people who through their own ignorance, thoughtlessness and misuse of their power, deny to others the opportunity of fulfilling the creative spirituality of which they are capable". The man who organised the lecture had to come over and ask him, ever so respectfully, to keep his voice down.'

Neville was stunned to silence. Jean had just summed up his unacknowledged feelings about his own upbringing, where his early chances had been intentionally scuppered by this father, and later opportunities made difficult for him because he wasn't connected into some old boys' network.

'I would have liked to have met him,' he said, genuinely moved. 'You've remembered word for word what he said about class.'

'He had a way with words. I liked to listen to him. I met him again on a walk across a meadow one Sunday,' said Jean. 'He had his paints and brushes in the pram. He let me paint with some of his materials. I did a small canvas of a five-bar gate. With cow parsley. He said it was "perfectly proportioned, and full of life and joy". I'll never forget that. It was the summer of the year he died.'

'You're a painter?'

'Only an amateur,' said Jean.

'Was this artist – Mr Spencer – was he helping you to realise your *creative spirituality*?'

'Yes. He wasn't interested in me in any other way, if that's what you mean. He already had two wives.'

'Two wives? At the same time?'

'I don't know. I don't think so. That was the local rumour. He was unconventional.'

'But he lived alone ... when you knew him?'

'Yes. In a small house in the village. Then he moved back to his boyhood home, a place called Fernlea. He had people living with him there. I certainly wasn't one of his visitors.'

Neville couldn't believe that he was experiencing a pang of jealousy over Jean's barely there relationship with an elderly artist, now dead. He felt compelled to secure her to himself somehow.

'If I ever have children … if I ever have children to bring up, I will make sure they have every opportunity. I would help them to go to university, not stand in their way. If they were able to pass exams and all of that, I would go and bang on the door of the university and I would say: "These are my children. *Do not deny them the opportunity of fulfilling the creativity of which they are capable.*"'

Jean was more than surprised at Neville's vehemence on this matter. She looked at him afresh, his face both indignant and passionate. She had dearly wanted to go to university, had been encouraged to do so by a teacher at Slough High School for Girls, but had to settle for less.

'I think so too,' she admitted. 'There must be many people who have been held back, like we have been, by the English class system just not allowing them the smallest chance of change. Not understanding the difficulties, choosing to put their head in the sand about the barriers that so many people experience – poverty, no guidance, class prejudice.'

Neville didn't notice that Jean was expressing her very first political realisation. He didn't want the conversation to start exploring generalities. He wanted it to be personal, about himself and Jean.

'Jean … I would like to see you again,' he faltered. 'Would you like that too?'

'I don't know. Let's wait. See if there are any other people on the Bureau's books for us to meet. You're the first person I've met through them. Am I the first one you've met?'

'Yes. But I don't want to go through all this again – I don't want to meet anyone else,' said Neville desperately.

The drive back to Cookham was wretchedly silent. The undeniable connection he had felt with Jean evaporated under her insistence that she wanted to try out more matches through the Marriage Bureau. She got out of the car, taking the picnic basket and rug. She thanked him quite warmly, but didn't invite him up to her flat for a cup of tea before his drive back to Cambridge.

On the way back, he thought he might be having a heart attack. The pain was severe, and he had to get out by the side of the road and breathe deeply until it had abated. He sat back in the driver's seat. Every part of him wanted to drive back the other way, to Cookham. He wanted to do anything other than drive back alone to his big, empty house.

Neville was at first relieved to see that his cleaning lady, Mrs Perkins, was in his house, but the next moment, he realised that she was in distress. Between noisy slurps of a cup of tea that he had made to calm her down, she managed to tell him what the trouble was. A man had knocked on the door, and when she opened it, he had taken a close-up photograph of her surprised face, said nothing at all, then got back in his car and had driven off.

'What sort of car was it?' asked Neville.

'Big, black,' said Mrs Perkins.

'Make, model?' asked Neville. Mrs Perkins looked blank.

'What did the man look like?'

'Tall, dark hair, dark glasses, a very big camera,' she said.

'He probably thinks that you're my wife,' said Neville with distaste, a little unfairly.

'What is going to happen? Why is he after you? What have you done? Are you wanted by the police?' asked Mrs Perkins.

'No, no, not the police,' soothed Neville. 'It's just the Russians. I don't think they are going to do anything to you. I think they are just warning me, letting me know that they are watching. They think, wrongly as it happens, that I am giving away their secrets to the Americans.'

'Secrets?'

Mrs Perkins was still shaken up, but the juiciness of this incident was almost too much excitement for her to bear. She couldn't wait to tell people that a Russian had taken her photograph.

'Have you been doing that? Something about secrets?' she asked again, wide-eyed.

'No, not at all. They've got it wrong,' said Neville, wanting to stamp out the beginnings of a rumour that could destabilise his life just as he needed to be perfect husband material for Jean.

'Look, Mrs Perkins, please don't worry about this. You're not in any danger, and I will deal with it. You can stop being my cleaning lady if you like. I'm soon going to have a wife of my own, anyway.'

Mrs Perkins's gossipometer needle was now off the scale.

'A wife?'

'I hope so. I don't know for sure. But I really hope so.'

'Is it the one who wrote to you? On that creamy writing paper with the lovely handwriting?'

'Sort of,' hedged Neville.

'I knew it!' shrieked Mrs Perkins. 'A love letter! You read it in your car that morning, and I knew it must be a love letter.'

Neville thought how wonderful, how utterly amazing it would be, to receive a real love letter from Jean. The thought then occurred to him that he could write one to her. But would she even want such a thing from him?

Mrs Perkins was up and putting on her coat, all thoughts of Russians gone from her head. The much more interesting matter of a potential wife, and a love letter, one that she had actually touched, were uppermost now. She had struck gold on the intrigue stakes, and she wasn't about to resign.

'I'll see you on Thursday,' she told him, as she rushed out of the door to start spreading rumours about a love letter and a wife.

City 40, and GCHQ

'Eric? It's Cocky,' said Neville that evening over the phone. 'The KGB have been round, giving me a warning. The neighbours are starting to notice.'

'Your phone line's probably bugged as well,' said Eric. 'Best if we can meet up. Look, I'll drive over. I'm only in London.'

Two hours later, Eric's car pulled up outside Neville's house.

'It's this advisory role, this GCHQ thing,' said Neville. 'Every time unexpected results come in from the test launches in Russia, or there's a new Soviet personality they think I might know, I have to drive to Cheltenham. I've used up all my sick leave and holiday. It doesn't let me have a private life, and now, it looks as if I'm being warned off by the KGB.'

'Look out for your brakes,' said Eric. 'It's one of their favourite tricks. Cutting brake cables – then making your resulting death look like a car accident. Are you sure this room isn't bugged?'

'No one has been in here since I moved in, except for Mrs Perkins, my cleaning lady.'

'I know from my sources', began Eric, 'that the Russians are hell-bent on getting a man into space.'

'They surprised everybody, particularly the Americans, when they were first with the Sputnik 1 satellite in 1957, carried into orbit on an R7,' said Neville.

'With a fully functioning *inertial* guidance system,' added Eric, with a raised eyebrow.

'Do you remember, in Gatow, when we tried to get them to stop focusing on weapons, by showing them evidence that the Americans were planning to be first in space?' asked Neville.

'The Russians were already designing a satellite for space exploration,' replied Eric, 'for the International Geophysical Year.[3] We probably made no difference.'

'It certainly felt ... decisive, when I handed Mitya those papers from that Anglo-American conference you'd been to at the Air Ministry. I felt like a spy. I could have been arrested by the KGB. And, I was stupid enough to criticise the Soviets' habit of casual murder – in a crowded café.'

'You had protection officers, in the café,' said Eric. 'I organised that much.'

'I could have lost my nerve,' said Neville.

'But you didn't. You transferred the papers to Mitya, he took them back to Moscow, and three weeks later, the Russian "decrees" had turned to space travel. I'm just saying it was probably coincidental. There was a lot going on.'

'Hmm,' said Neville.

'And now, after what happened to that poor dog,[4] they want to get a human being into orbit, in a sort of cockpit. They are aiming to be ready next year.'

'The Americans are planning something similar,' said Neville. 'But GCHQ are uneasy. They don't want to get diverted from the realities of the Cold War. They know that Khrushchev is still stockpiling ICBMs.'[5]

'Khrushchev gave that famous speech in 1956,' replied Eric, 'the one where he denounced Stalin. "The Personality Cult and its Consequences", about Stalin's crimes against loyal party members on false charges. The failings of his agricultural policy, causing starvation for so many people. Ordering mass terror against dissenters. But mostly, for insisting on adoration from ordinary Russian people, as if he was their god.'

'I thought, after that speech, Khrushchev would publicly abandon ICBM development,' responded Neville. 'To set a peaceful example. To be the opposite of Stalin. It's so disappointing that he didn't.'

'There are still arsenals of fearsome weapons being built up on both sides. The public negotiations and agreements that have been made on the world stage are being quietly flouted. There's still no genuine trust,' said Eric.

'The space race certainly is a diversion,' agreed Neville. 'Did you know that they've finally recognised Korolev, publicly, as "the Father of Space Exploration" – as if that was what he was doing all along?'

'He could hardly be celebrated as "the Father of Mass Destruction by Nuclear Weapons",' said Neville. 'But thinking about Korolev, especially knowing that he was one of the loyal party members unjustly imprisoned in that gold mine by Stalin – he never actually wanted to develop weapons at all. Mitya told me that at a critical moment, to postpone the launch of the R7,[6] he authorised the building of Baikonur.[7] Huge scale, with massive radio towers, which took up a huge chunk of budget and two whole years. He was *still* doing all he could to slow things down.'

'A man of peace, indeed,' said Eric. 'It's possible. I often think of that bit in the Bible:[8] "Blessed are the meek, for they will inherit the Earth".'

'"Blessed are the meek, for they will have an earth to Inherit",' said Neville. 'Adapted from God's word by me. The more I think about it, Korolev and many more were trying to be peacemakers.'

'In the West, he would be celebrated for that, given a Nobel Peace Prize for working as hard as he could towards a peaceful resolution, at great risk to himself, in the full glare of the Kremlin,' said Eric.

'His name isn't even widely known in the West,' said Cocky. 'As a peacemaker or anything else.'

'And yet, under Korolev's leadership, the Soviets went from nought to Sputnik in just twelve years,' smiled Eric. 'With the tiniest bit of help.'

Eric and Cocky had never acknowledged their collusion via Cocky's homemade radio on Gorodomlya. It was a long-ago world, with very different pressures from their present one. 'I'm just glad I never took Kuznetsov[9] through it all,' said Cocky. 'I got ill just at the right moment.'

'Then, you escaped,' said Eric.

'Again, that was pure luck, that quartet being stuck on the island.'

'Didn't you once say that you invited them?' queried Eric.

'I couldn't do the inviting, but I identified the quartet. I wanted outsiders to come, anyone but Russians,' said Neville. 'Finnish people fitted the bill. I was hoping they could alert our government to my situation on Gorodomlya. The Germans were being repatriated slowly. I was forgotten, the only Brit, and I'd been using that secret radio to communicate with you. I thought the Finns might alert Churchill, to prevent me from being "disappeared".'

'You thought our government could take action – but it couldn't have done. Getting out of the situation in the way it happened was your best option,' said Eric.

'I still have nightmares about that shaky escape plan,' said Neville. 'It had to work, but we made so many mistakes. I have a recurring nightmare about not being able to play a violin when my life depends on it.'

'Violin?' queried Eric.

'Yes, at Vaalimaa,' said Neville, aware that Eric had forgotten details that were etched into his own memory. 'I can hardly believe I'm here, chatting to you in a comfortable house in Cambridge.'

Eric refilled their glasses, to emphasise the relative sophistication of their present circumstances. 'We need to ignore the crazy space stuff, just let it roll,' he said eventually. 'GCHQ is more concerned with the nuclear threat.'

'The Americans have got Hanford,' said Neville. 'They were the first to build a city around a plutonium production plant, even before the end of the war. I shouldn't be surprised if the Soviets copied them.'

'Do you remember at Gatow, the reconnaissance pilots came back with footage of that closed city built around a plutonium plant in the Urals? City 40?'

'Yes. I remember,' said Neville. 'Like something out of a science fiction story.'

'It caused real consternation in America,' said Eric. 'They became convinced there were many more Russian cities like it. Dedicated to nuclear bomb production. It ratcheted up the tension.'

'Like bigger versions of Gorodomlya. Workers having to commit to something wanted only by the Kremlin,' said Neville.

'On that subject, I've been wondering what to do about an intriguing letter that was sent to Churchill at the Air Ministry about a year ago,' said Eric. 'Churchill passed it on to GCHQ, and they've passed it to me. It had taken months to reach London, and it has been really heavily censored. The postmark is Yekaterinburg in the USSR. I'll just get it from the car.'

Eric went outside to his car, opened the boot for his briefcase and brought it inside. He pulled out a much-thumbed letter, just two sheets, but clearly, it had been three sheets before one complete sheet had been removed. It was written on poor-quality paper with a pencil that could hardly make a mark.

Eric looked through it, as if he didn't know its contents word for word. 'It doesn't mention plutonium as such,' he said, 'but that would have been blacked out by the censors. It does mention "all the components you need to build a radio". I've been meaning to show it to you. It mentions an island. It's obviously code. But what does it mean? And, the writer says he's dying of leukemia. Could that be code as well?'

Neville took the letter eagerly. He scanned the handwriting – so illegible and badly formed and yet so familiar. Recognising the handwriting felt like a punch to his stomach.

'Walter,' he breathed.

'Who is Walter?'

'Wait a minute,' said Neville. 'This is from an old colleague of mine on Gorodomlya. Walter. He was "disappeared" by the Russians. He may have died from leukemia by now, if you've had the letter for a year. The radio he mentions – it's his way of telling me the letter is from him. He says he's got *all the components* he needs, in this closed city. He says the city is *like the island*. He says he works there, not in the *something* plant, but in the site communications hut. I think he's managed to send the letter from there. He's telling us ... trying to tell us ... that it's a plutonium production plant. The censors have blacked out the name, but he must have somehow got hold of it again before posting, because he's written "Chelyabinsk, City 40". He's scribbled something as well. I think it's "not on map".'[10]

'Oh. I thought that said "got a mop",' said Eric. 'The writing's so hard to make out.'

Neville was reading the letter with too much concentration to laugh at Eric's joke.

'How do you know it's from this Walter?' said Eric. 'There's no name, no signature. And why didn't the Soviet censors in the place he sent it from react more to it being addressed to Churchill? Why didn't they confiscate the whole thing?'

'It's definitely from Walter. It's his uniquely bad handwriting. He's the one who helped me build my radio. He stole the components for me, from Gröttrup's workshop ... when he was more trusted than I was. He helped me build the radio I made for contacting you. He was ... "disappeared". Now I know he was sent to this place, City 40. Walter has just written "Winston Churchill" on the envelope, no indication he might be the prime minister.'

'He's not prime minister anymore,' said Eric. It was Clement Atlee till 1951, then Churchill again till '55, then someone else for a couple of years. Now it's Harold Macmillan.'

'Walter wouldn't know all that, though, would he?' snapped Neville. 'He's in a bloody closed city with thousands of other forced labourers, building a highly radioactive, *dangerous* nuclear plant in the middle of nowhere. No movement or communications in or out, except, I suppose, for the essential communications needed to manage the work. He's got himself into the only possible place he could send a message from. He's an absolute genius.'

'I should have shown you the letter earlier,' said Eric. 'How could I have known it was meant for you? Walter was probably protecting you by keeping your name out of it.'

'He's dying, or has already died, from leukemia,' said Neville, his anger mixed with admiration. 'I can hardly imagine what he must have gone through, just to write and post this letter.'

'I'm going to get on to GCHQ first thing tomorrow,' said Eric. 'And, Gatow will need to send another photographic mission for new aerial pictures of City 40.'

'I need another whisky,' said Neville. 'Getting this letter, brilliant though it is, is a total shock. Would you like a whisky? Do you want to stay over in my spare room?'

'No thanks, I'm getting back to Marianna and the children,' said Eric. 'She doesn't like to be without me; she's had years of me being away.'

But Eric had to stay over in Neville's spartan spare room after all. While the two friends had been examining Walter's letter, somebody had cut the brake cables on both of their cars. Eric came back inside to report the fact he tried his brakes, as he always did before setting off, and they had stopped working.

The next morning, when Neville went outside to repair both sets of brake cables, he noticed the cleanly cut-through wires.

'You might have to move,' said Eric to Neville over scrambled eggs. 'The KGB are obviously going to keep going with their dirty tricks. Go to ground. Merge in. Change your identity. Get married or something.'

Chapter 7

Escape to Canada

Cold War Marriage

'I'd love to get married,' said Neville. 'I really would.'
'Have you met anyone yet?' asked Eric, fully expecting a negative reply.
'Yes. I don't know. I hope so.'
'Whoa, that's a turn-up!' said Eric. 'What's her name?'
'I'm not sure yet. I only met her yesterday.'
'You're not sure of her name?'
'Of course I know her name. It's Jean.'
'Jean and Neville, Neville and Jean, true love forever,' teased Eric. 'Jean and Neville sitting in a tree, K-i-s-s-i-n-g,' he sang in a childish voice.
'If only I could be sure of her,' began Neville earnestly. 'Yesterday, we …'
'Marry her, marry Jean, Neville. Move away from here, get another job, and get away from the KGB.'
'I can't ask her to marry me. She doesn't know me well enough yet. She wants to go out with other people first. I'm scared, Eric, that if I ask her, she will say no.'
'I felt the same way about Marianna,' said Eric seriously. 'I couldn't believe that she actually liked me enough to say yes. You've just got to take your courage in both hands.'
'She won't say yes,' replied Neville, standing up to finish his cup of tea before setting off to work in his newly mended car. He was going to be late.

***

My parents, Neville and Jean, married the following year, in the summer of 1961. It was a small church wedding, with the reception held in the corner-plot garden of Jean's parents' council house in Slough. On the morning of the wedding, Jean's friend from schooldays asked her how it felt to love someone, and Jean replied that she didn't know, but she hoped to find out.

* * *

Neville had asked Jean's parents for her hand in marriage. They had not objected, either to the ten-year age difference or to her lack of enthusiasm. My grandma, Jean's mother, observed that if she was to have her children before she was 30, she should get started.

A few years later, their own 'cold war' began.

There were unexplained absences on Neville's part that did nothing to build their fragile trust. He would disappear for days at a time, and he wouldn't phone home. Sometimes he would mention 'working in Cheltenham'; other trips were to Germany, the USA and Canada.

Jean tried to get to know people from Neville's various workplaces, so she could ask them where it was that he went. Neville didn't invite Jean to any of his work events, in case she found out the trips weren't to do with work. He told his workplaces the trips were family holidays. He was fiercely loyal to the Official Secrets Act.

Jean was lonely and bored, bringing up three children (born 1962, 1963 and 1964) single-handedly and trying with varying degrees of success to do the one thing that Neville asked for – to have an evening meal ready at six o'clock.

Neville spent huge tracts of time in the garage tinkering with the car, or in his study with his headphones on, listening to classical music.

As a small, observant child, I understood that my father squarely equated meals with love. Every time there wasn't one, it wasn't ready, or was burnt, he retreated to his study. It was more than annoyance at Jean not providing the *one thing* he wanted: he would feel disrespected.

My parents' tense stand-off lasted ten years. Misunderstanding of signals, distrust, miscommunication, and failing to discuss each others' fears, combined with a fundamentally different world view. Their misjudged alliance led to significant damage to innocent people. All the elements of the Cold War were still playing out in the background of their lives.

The end came quietly, without an explosion. Just a confusing, destabilising absence.

Epilogue

As I write in January 2024, a newspaper headline shouts 'Hostile threat highest since the Cold War'.

According to reporter Martin Evans, Britain is facing an acute threat owing to the triple menace of Iran, China and Russia. Of the three, the most urgent threat, he reports, is Russia, with Germany warning that it is gearing up to attack NATO countries *within the next five years*.

The extreme weapons, the communications, the satellite capability, the ideological divisions, and the eminence of paranoid, even psychopathic, rulers don't bode well for peace.

My grandfather, Iggy in the book, was a destroyer, a fraudster, a user of others, and a supremely selfish man. He managed to create a living hell for his family, just as world leaders can create living hells for their populations.

These men (presently, the warmongers are all male) can't see that in order to be Great, they need to build – build connections, find common ground, put aside anger, risk misunderstanding, defuse tension, and be mindful of their actions. Seeking peace is strenuous and difficult. For the truly great men and women who are capable of leading the world in this way, every generation will study them, and emulate their powerfully peaceful ways.

List of People in the Gorodomlya Island Project

Ackermann, Eric George (1919–86)
Eric was born on the Isle of Wight, and brought up in London. He was involved in flying missions to intercept enemy radar and communications, earning the George Medal. From 1945 to 1959, he developed the RAF's land-based radar intercept network in Germany. He was Commanding Officer of Air Scientific Research Unit (ASRU), which became 646 Signals Unit in 1952. He established direct secure communications with the newly created American CIA in 1947. He left the RAF in 1960, moving to the USA, where he worked on satellite communications until retirement. Eric died aged 66.

Ackermann, Marianna (1924–94)
Gizella Maria Anna von Schmidt, Eric's second wife, married in 1948. Met in Salzburg (or could have been Ried) on 8 February 1946. Marianna was a Hungarian refugee who fled the Red Army with her sister in 1945. Eric and Marianna had two sons, Peter, born in 1952, and Nicholas (Nick), born 1957.

Appleton, Sir Edward (1892–1965)
British scientist, winner of Nobel Prize for Physics in 1947 for his contributions to the knowledge of the ionosphere, which led to the development of radar. Member of the Department of Scientific and Industrial Research and the War Cabinet's Scientific Advisory Committee.

Churchill, Sir Winston (1874–1965)
Sir Winston Leonard Spencer Churchill served as Prime Minister of the United Kingdom from 1940 to 1945, and 1951 to 1955. From 1945 to 1951, he was active as Leader of the Opposition, and held post-war roles, such as

involvement with the Cold War. Churchill coined the term 'iron curtain' for the designated line of influence of the Soviets.

Cox, Bruce Neville (1925–2024)
Author's father and 'Cocky' in the book. Born in Bristol, he was one of six surviving children. At age 14, he left school to work on aircraft design at Bristol Aeroplane Company, Filton Works. Educated as an engineer and draughtsman through National Certificate classes in the evenings. Joined the RAF in 1946, service number 4017382. Worked on the inertial (gyroscopic) guidance system and analogue computer for rockets. In Nova Scotia, Canada, he worked with buoy instrumentation. Held one patent 'to do with the manufacture of microchip resistors'.

Married Jean Ridley in 1961; three children born 1962, 1963 and 1964. Emigrated to Canada in 1972. During his final days, he lived in a care home in Canada.

Bruce Neville Cox died in April 2024.

Dieminger, Dr Walter (1907–2000)
Distinguished German space scientist. Headed the Fraunhofer Institute for Ionospheric Research until 1946. When these laboratories were transferred to Lindau, they eventually formed part of the Max Planck Institute for Solar Systems Research. Dr Dieminger was the first director until his retirement in 1971.

Engelbach, Paddy (1922–55)
Patrick Stevens Engelbach worked as a signals interceptor at 646 Signals Unit in 1952. Patrick had married Helena Verna Philby in September 1946. Helena was the sister of spy for the Soviets, Kim Philby, which was not known at the time. Paddy died in 1955, in an air accident at RAF Coltishall, Norfolk, aged 33.

Gröttrup, Helmut (1916–81)
Former deputy to Wernher von Braun, whose team designed the V-2 rocket bomb at Peenemünde. Husband to Irmgard and father to Peter and Ursula. On Gorodomlya, he designed the R14 rocket that replaced Korolev's R3.

Tasked with developing an 'electronic computer' in the final months before being repatriated from the island in 1953. Debriefed by the CIA in 1957, but downplayed the German work. Post-Gorodomlya, in 1967 in West Germany, Gröttrup filed patents for incorporating an integrated circuit chip onto a plastic card, which contributed to the invention of the smart card. He remarried in 1964.

Gröttrup, Irmgard (née Rohe)
Married to Helmut 1940–63. Two children, Peter and Ursula (Ulli). Community leader, later wrote *Rocket Wife*, published in 1959. Divorced from Helmut in 1963.

Gröttrup, Ursula (Ulli)
A child in the book, Ursula became a professor of psychology and journalist in Germany. She wrote '*Wissenschaftliches arbeiten unter Hitler und Stalin*', about her father Helmut's scientific legacy, presenting it as an illustrated talk at the Deutsches Museum, Munich, in 2017.

Jones, Dr Reginald Vincent (R.V.) (1911–97)
British physicist and scientific military intelligence expert. Served in the Air Ministry 1936–46, as Assistant Director of Intelligence (Science), reporting to Churchill, after which he was Chair of Natural Philosophy at Aberdeen University. Involved in scientific intelligence throughout the Second World War, he became known as the 'father of scientific and technical intelligence'.

Korolev, Sergei (1907–66)
Sergei Pavlovich Korolev was born in Zhytomyr, Ukraine. He was a leading Soviet rocket engineer and spacecraft designer, credited posthumously with the development of the Sputnik 1 satellite and its carrier rocket. Korolev was held in a Soviet gulag in Siberia from 1939 to 1940 and in a different prison camp until 1946. He was 'rehabilitated' and made Chief Engineer of the Nordhausen Institute in Germany, with Gröttrup, in 1946. Considered posthumously to be the father of space exploration. The present city of Podlipki/Kaliningrad, is now named Korolyov in his honour. He died aged 59.

Hölzer (Hoelzer), Helmut (1912–96)

Scientist in Nazi Germany who developed the first electronic analogue computer. Based on an electronic integrator and differentiator, it was used in the V-2 rocket. At the beginning of 1942, he and his team developed the Messina telemetry system to track position and orientation of a rocket body. He went to America in 1945 as part of Operation Paperclip, where he continued work on the computer, which is essential to inertial navigation.

Kuznetsov, Viktor Ivanovich (1913–91)

Appointed head of N11-10 institute in 1946, dealing with gyroscopes. Colleague of Korolev. Developed a theory of gyroscopes 'from scratch'. Awarded Order of Lenin in 1960 for 'creation of high-precision devices'. Received the State Prize of the USSR in 1943 and 1946. Developed a number of unique instruments and systems. Citation: the *Great Soviet Encyclopedia*, 3rd edition.

Mishin, Vasily (1917–2001)

May have been 'Mitya'. Prominent Soviet rocket scientist. Succeeded Korolev as chief designer in 1966. Worked on N1 rocket programme intended to land a man on the Moon.

Magnus, Dr Kurt (1912–2003)

German rocket scientist, expert in modern navigation technology and inertial sensors. Gyroscopes expert. Drew a map of Gorodomlya Island from memory by hand.

Piggott, Dr William Roy, OBE (1914–2008)

International British leader in ionospheric research and assistant to Sir Edward Appleton. Received the OBE for establishing ionospheric research at Lindau in 1946. Head of Atmospheric Sciences at British Antarctic Survey. Edited the still valid rules for ionospheric soundings. Known for his 'mad professor' looks and mannerisms.

List of People in the Gorodomlya Island Project   223

Preikschat, Fritz Karl (1910–94)
Workshop manager for N11-88 on Gorodomlya Island from 1946 to 1951. Engineer and head of the radio-controlled guidance system workshop. Released in June 1952, he spent two months being debriefed by the US Army on the Soviet Union's rocket programme. Wrote a report on the Soviet Union's 'Microwave-based control system for long-distance rockets', published in 1954.

Rautavaara, Einojuhani (1928–2016)
Born in Helsinki, studied musical composition at the Sibelius Academy, 1948–52. Also studied at the Juilliard School, New York. Wrote String Quartet No. 1 in 1952, rewritten in 1953 as a quintet. Wrote String Quartet No. 2, *Unknown Heavens*, in 1958, and *A Requiem in Our Time* in 1954.

Spencer, Stanley (1891–1959)
Renowned modern artist. Lived in Cookham, Berkshire, in his boyhood and old age. Eccentric presence who wheeled his painting materials in a dilapidated old pram.

Stalin, Josef (1879–1953)
Georgian-born successor to Lenin as Soviet leader. Architect of disastrous agricultural policy that led to mass starvation. From 1945 to 1953, under his rule, more than 750,000 Soviet citizens were arrested by his police force, the NKVD (precursor to KGB), and punished on 'suspicion of espionage'.

Starling, Jennetta (née Ridley, was Cox) (b. 1935)
Author's mother and 'Jean' in the book. Amateur artist and teacher. Lives near London.

Ustinov, Dmitry (1908–84)
Marshal of the Soviet Union during the Cold War. In October 1949, he led a delegation to Gorodomlya. In March 1953, on Stalin's death, he became Soviet Minister of Armaments.

List of Places in the Gorodomlya Island Project

Air Ministry Headquarters, London
Department of the UK government, responsible for managing the Royal Air Force. In 1945 known as Adastral House after the RAF motto *Per adua ad astra* (Through adversity to the stars). In 1955, the building was used by the BBC as Television House. From 2013, it was acquired by China Overseas Holdings Ltd., and is today known as 61 Aldwych.

Bristol Aeroplane Company, Filton Works, Bristol
Founded in 1910, the company's first premises were former tram sheds at Filton. In 1959, a merger with several major British aircraft companies formed the British Aircraft Corporation (BAC). Now BAE Systems, the company still has a presence at Filton.

The Brocken
Highest peak in the Harz Mountain range in Northern Germany. A narrow gauge steam railway, in place since 1899, still takes visitors to the top. From 1945 to 1947, occupied by US troops; in 1947, transferred to Soviet territory, which became the border zone of East Germany.

The Brocken was used extensively for surveillance and espionage through two large listening stations on its summit. Today, FM radio and TV broadcasting, and tourists, make use of the Brocken.

RAF Changi, Singapore
Changi was the terminus for the main RAF transport services to the Far East. The main period of building from 1948 to 1956 included headquarters, hangars and the runway itself, all carved out of jungle and swampland. In 1972,

the RAF withdrew from Singapore. The site is now Singapore International Airport.

Fraunhofer Institute for Ionospheric Research, Ried im Innkreis, Austria
Walter Dieminger founded the Luftwaffe's Centre for Radio Transmission in 1943. It united with the Fraunhofer Institute for Ionospheric Research at Ried im Innkreis, in 1944. Relocation to Lindau was completed by the end of March 1946. It became part of the Max Planck Society, and was renamed the Max Planck Institute for Ionospheric Research.
   Fraunhofer-Gesellschaft remains Europe's largest and best-funded scientific research organisation, with branches across the world.

RAF Gatow, West Berlin, Germany
In 1951, the location of this RAF base became important for intelligence gathering by RAF signals experts monitoring Soviet air traffic broadcasts from bases in Eastern Europe. Chipmunk photographic reconnaissance planes flew from here from 1954. Now it is home to the Luftwaffenmuseum der Bundeswehr, or German Air Force Museum.

Gorodomlya Island, Tver Oblast, Lake Seliger, Russia
Once a monastic island, in 1928 the Soviet government ejected the monks and turned the site into a biological research institute. In 1940, 6,000 prisoners of war were kept here before execution in nearby Tver.
   From 1947 until 1953, about 40 out of a total of 170 German rocket scientists and engineers, including families, from the Nordhausen Institute in Germany, were forcibly moved to the island. The Soviet designation for the island was Branch 1 of N11-88. In 1952–3, apartments and facilities for Russian rocket workers were built. Today, the settlement has 3,800 residents.
   Lake Seliger today is a popular tourist destination with sandy beaches and woodland walks. It freezes over annually to the point where vehicles can drive across it. The institute building is now part of Zvezda, a subsidiary of Roscosmos, the State Corporation for Space Activities.

Ostashkov
The Russian provincial town nearest to Gorodomlya Island. Its architecture and setting by Lake Seliger make it a popular resort town for visitors.

Obernkirchen/646 Signals Unit/Air Scientific Research Unit (ASRU)
RAF Obernkirchen is 7km east of Bückeburg in Northern Germany. Operational from 1945 to 1955. In 1949, it became the Air Scientific Research Unit, dedicated to keeping track of Soviet military scientific advancements and activity. In 1952, the name changed to 646 Signals Unit. In March 2009, records of its activities were destroyed due to asbestos contamination and water damage.

Ozyorsk/City 40/Chelyabinsk-40/Mayak Plutonium Production Site
Founded in 1945, a specially built city where plutonium is processed for nuclear weapons. Sited near Yekaterinburg, 1,500km from Moscow. A nuclear explosion incident in 1957 was covered up until 1980. People were given jobs 'for life' but not allowed to leave. Radioactive waste dumped in Lake Irtyash caused total death to the lake. Three times the radioactivity level of Chernobyl.

Vaalimaa border crossing, Finland
A border checkpoint since 1917, a border crossing station for road traffic was built in 1958. It is known for very long traffic queues, mainly due to security measures on the Russian side. Since Finland was a neutral country in the Cold War, it did not protect illegal border crossers, who were returned to the Soviet Union if captured.

Acknowledgements

Penrose Halson, author of *Marriages Are Made in Bond Street – true stories from a 1940s marriage bureau*, was so incredibly helpful when I approached her for details of my parents' meeting through Heather Jenner's Marriage Bureau that she became a valuable reader and critic of my manuscript. Penrose ran the Katharine Allen Marriage and Advice Bureau and is fascinated by stories of marriage bureau couples such as my parents.

Chris Shurety, MBE has been a reader and advisor on the manuscript, as well as a general source of encouragement, advising me to 'just get it published' on more than one occasion.

Bruce Neville Cox, despite the surprise and upheaval of not just connecting with me after fifty years, but also working on this book, kept calm and patiently explained many things.

David John Cox, my younger brother, for suggesting in the first place that I join him in regular Zoom calls. And for being supportive of the project throughout.

Nick Ackermann, second son of Eric, who has provided detailed information, much useful editorial advice, and encouragement for the project, from his home in America.

David Haysom, co-author of *Covert Radar and Signals Interception*, and himself a retired intelligence officer, provided photographs and context for the Air Scientific Research Unit.

Jennetta Starling, 'Jean' in the book, and my mother. Acknowledging her support and help, particularly her honesty over how she met my father.

Charles Vickers, my husband, always ready with a hug and a cup of tea when the writing became overwhelming at times. And our two adult children – a scientist and a writer – for your support.

Leeds University Rocketry Association members **Cem Mirata** and **Theo Youdis** for their valuable input to the rocket science in the book.

Endnotes

Prologue

1. Peenemünde was a German rocket research and testing site on Usedom Island, on the Baltic coast. It is where the world's first V-2 rocket was developed. The British attacked it from the air in 1943.

Chapter 1: Biographical Notes

1. Luftwaffe means 'air weapon'. The German equivalent of the UK's Royal Air Force (RAF).

Chapter 2: Joining the RAF

1. John, christened Basil. The siblings' names began with B, but most switched to a middle name.
2. The Fraunhofer Institute for Ionospheric Research, which once had 300 staff, was only based at Ried im Innkreis in Austria for a couple of years, from 1944 to 1946.
3. Reginald Victor Jones, government physicist tasked with investigating the German application of science to air warfare. Considered to be the 'father of signals intelligence'.
4. In Chertok's *Rockets and People Vol. 1*, written in Russian in 1994, he states that Korolev was 'recalled to Moscow in early February 1946, returning to Bleicherode in March'. These are the dates of the Fraunhofer mission, where my father remembers Korolev, for at least part of that time, being in Ried, Austria.
5. Union of Soviet Socialist Republics (USSR).
6. In May 1946, the name 'Bleicherode-Nordhausen' changed again to Mittelwerk; in October 1946, the whole concern was forcibly moved to Russia by train, in Operation Osoaviakhim.
7. Germany had been defeated. Some German scientists were 'missile specialists', but were without jobs, and without power.

8. Missile development workshops from the Nazi era, operated in disguise as scientific institutes.
9. The study of modern 'rocket science' was very new; only the Peenemünde missile scientists could be considered pioneers. Most others, including Korolev, came from the aircraft industry.
10. It is interesting to note that Roy Piggott, who had been a student of Sir Edward Appleton, eventually received an OBE (in 1954) for the success of this mission.
11. Winston Churchill had been defeated as prime minister in 1945, and was at this time, in early 1946, Leader of the Opposition. He became Conservative prime minister again from 1951 to 1955.
12. Defeated Germany was divided into 'zones' occupied by each of the Allies – America, Britain, France and Russia.
13. Before the war, Walter Dieminger had established an ionospheric observation station at the Luftwaffe test grounds in Rechlin, Mecklenburg. Dieminger's relocation to Ried im Innkreis in 1944 was a move to civilise his work, away from military uses.
14. Years later, my father said that Eric completely underestimated his ability to absorb such a lot of technical and military information. In his turn, my father was too proud to slow Eric down to a point where he might be able to take it in. He was out of his depth, and prevented from humility by his own need to project his intelligence.
15. We now know the atmosphere has five major and several secondary layers.
16. Edward Appleton had made first proof of the existence of the ionosphere in 1924. For a while, it was called the Appleton layer and is now known as the F layer.
17. UK legislation that provides for the protection of state secrets. The OSA 1911 was in force at this time; now it is OSA 1989.
18. Tubular metal structures used for miniature research, or sounding rockets. Sounding rockets based on the V-2 were used at this laboratory, to study unexplored features of the upper atmosphere.
19. Military radio and radar signals.
20. Winged bombs powered by a jet engine that 'screamed' until it cut out. Known as 'doodlebugs'.
21. Piloted bombers were perilous. In the Second World War, only 24 per cent of British bomber pilots survived.
22. V stands for '*Vergeltung*', or vengeance weapons.
23. As a young man, my father often covered nervousness and embarrassment with humour. He thinks he did the impression of the doodlebug because he was

coming over to Eric as a know-all about V-1s and V-2s, when Eric knew so much more than he did.
24. Eric had flown in bombing raids to destroy V-1 launch pads in France in 1943, called Operation Crossbow.
25. In 1946, the Soviets were still officially allies of the British and Americans.
26. My father realised first-hand on this journey that European towns and cities other than London had been devastated, and that the populations of Germany and Austria had more in common with the ordinary people of London then they did with their national war leaders.
27. As I researched this episode, it struck me that Eric Ackermann and my father had much in common. Their resulting friendship, despite different ranks, was apparent. My father was never gregarious, and he didin't claim close friendship with Eric. He preferred to give his attention to cars, electronics and machines. That he and Eric took regular 'spins' in various models of car a few years later, when they were both at the ASRU in Obernkirchen, was, according to my father, all about the cars.
28. The Bruneval Raid shows the purpose and value of getting hold of enemy equipment. This raid gave knowledge of the state of German radar technology.
29. Many years later, Eric's son Nick tells me that his father met his mother, Marianna, at Ried im Innkreis, at 8 a.m. on 8 February. My father's memory was that although he slept through the Salzburg stop, rumours went around that two Hungarian sisters had been picked up there on the way, and that one of these turned out to be Marianna. It's possible that both versions are true – Eric may not have met Marianna properly until she was settled as a domestic helper at Ried.
30. The post-war economy of Germany was such that scientific jobs could no longer be supported. Most faced unemployment. It's possible that if Dr Dieminger had not received the British government's offer to relocate his laboratory to Lindau (in the British zone of occupation), he might have considered the Soviet's offer of a post in their rocket development programme.
31. Dr Dieminger, despite his commitment to peaceful scientific studies, and the warmth and humanity of his personality, had been head of the Luftwaffe test centre at Rechlin, Mecklenburg, from 1934 onwards. Following Germany's defeat, Dieminger would have been aware of the crumbling alliance between the Allies and Russia. He knew about the enormous V-2 rocket and weapon facilities at Bleicherode-Nordhausen, because some of his former colleagues had gone to work there. His own lifeline, to relocate his laboratory to Lindau, came from the British government, possibly through Sir Edward Appleton.

32. Dieminger knew that Russian scientists were currently combing out laboratories looking for equipment and evidence of V-2 experiments.
33. The Russian vehicles may have been Studebaker US6s, or ordinary Soviet army trucks. Studebakers had been sent to Russia in 1941 as part of 'Lend-Lease' agreements between the Allies.
34. Korolev had been reinstated after nearly six years (1938–45) of being a badly treated forced worker in a goldmine, a prisoner of Stalin's dictatorship. He wasn't at the top of his scientific powers, but he was under extreme pressure to deliver results.
35. William Roy Piggott, a student of Sir Edward Appleton, looked like a stereotypical absent-minded professor.
36. After the war, the race to miniaturise gyroscopes for weapons navigation systems resulted in midget gyroscopes that weighed 85g and had a diameter of 2.5cm.
37. These were sounding rockets based on the V-2.
38. Ethanol is a flammable gas that shouldn't be stored in this way.
39. Russians like to freeze their vodka, or make a slushy consistency by adding water or juice.
40. A lateral accelerometer measures the change in velocity of a moving body, specifically its cornering forces. It detects and measures turning.
41. Dieminger later became director of the Max Planck Institute for Aeronomy 1955–75.
42. Cosmic static is a key topic, as it is one reason why radio-controlled rocket guidance wasn't reliable – a matter that becomes important later. Cosmic static was later proved not to emanate only from the sun, however, but from background radiation of space, a residual effect of the Big Bang.
43. Jamming is intentionally sending radio signals on exactly the same frequency as the enemy's, and effectively neutralising them. Used to wipe out communication at critical moments, it had been responsible for blunting air attacks. Both sides had found ways to prevent their equipment from being jammed, and both sides had developed ways around the protective devices.
44. The dialogue here is fictional, introduced to convey the unintended consequence of Cocky's know-all way of explaining things. It makes people feel small, and they stop co-operating.
45. VHF band is still used for DAB and FM radio.
46. Thermionic valve for signal amplification.
47. At this time, homosexuality was illegal in the RAF and other services, and could be met with severe punishments. At the same time, it was generally too

much trouble to bother punishing people who were discreet, and only vindictive senior staff went through the process. Wappo and Freddie are fictional, but they illustrate a known Services phenomenon.
48. This trick was designed to pay the Russians back for an earlier, similar deception (Blizna, 1944) when they had substituted rusty old truck and tank parts for V-2 rocket parts en route to Britain.
49. RABE stands for *Raketenbau und Entwicklung*, an acronym that also means 'Raven' in German. In February 1946, the name changed to Institute Nordhausen. In May, it changed again to Mittelwerk, or Central Works.
50. In reality, my father made some of these connections in understanding of the situation he had glimpsed, well after the event, in the years following. He didn't interpret the 'abduction trains' drawing on the back of the Charlie Parker album until after the mass movement of ninety trains from Nordhausen in October 1946.
51. Operation Osoaviakhim copied the American Operation Paperclip. In the Soviet's version, 170 specialists from Mittelwerk, plus families, were part of a mass abduction of over 2,300 Germans. The Mittelwerk Germans were taken to N11-88 in Podlipki near Moscow; and Gorodomlya Island.
52. From our modern vantage point, where we accept that huge numbers of NASA and other scientists with massive national budgets routinely explore space, the idea of space travel was once unimaginable. It was a dream, not a national priority. Space travel is indeed a strange use for national budgets decimated by expenditure on war.
53. Wife Ksenia and daughter Natasha.
54. The identity of Mitya is far from secure. He may be Vasily Pavlovich Mishin, who took over from Korolev as chief design engineer of the Soviet Union's moon exploration programme, when Korolev died in 1966. Ten years younger than Korolev, he could conceivably have held a 'son-like' place in his affections. Mishin is remembered for being at the helm when the Americans beat the Soviets to land a man on the Moon. An article in the *New York Times* described Mishin as 'a solid engineer lacking in his boss's charisma'. Mitya, of course, may be someone else entirely.

Chapter 3: Air Scientific Research Unit, Obernkirchen, 1951

1. HMT *Empire Windrush* continued as a troopship until 1954. It is better known as the vessel that transported about 800 passengers from the West Indies to Tilbury Harbour in the UK in 1948.

2. Radio beams to guide aeroplanes in enemy airspace, a precursor to radar.
3. Eric at this stage was working closely with American-based scientists from MIT (Massachusetts Institute of Technology) to find alternatives for radio guidance control systems. Alternatives were being sought because radio control systems were easy to jam deliberately or due to cosmic static. Efforts to develop an inertial system were just beginning. Eric had a rudimentary inertial guidance set-up in his office, made with components of a 'closed loop guidance system' he had salvaged from a very early V-2 fragment.
4. Wernher von Braun led the design and development of the original V-2 rocket at Peenemünde.
5. This version of events is corroborated by Irmgard Gröttrup.
6. Paddy Engelbach.
7. Sgt Dickie Hunt.
8. Geoff Easthough.
9. Whilst my father and everyone else took the Official Secrets Act very seriously, and everyone had read a copy of it before signing it during their careers, there was often real confusion about what information was or wasn't to be passed on. People veered from obsessive adherence to almost disregarding it, because it was difficult to interpret when faced with real-life scenarios, especially in a place like this, which relied on open knowledge sharing.
10. The test launch in October 1947, one year after the mass abduction of scientists in October 1946, was the first where Helmut Gröttrup supervised the process of the launch at Kapustin Yar. Unknown to Korolev or Gröttrup at the time was the fact that the radar signals emitted during and after launch were picked up and relayed by the ASRU to Britain and the USA. The rocket flew 206km before disintegrating and landing 30km off-target. A second launch, three days later, went 231km from the launch site in the wrong direction due to control systems failure.
11. That important deductions were frequently made using a combination of experience with other communications, guesswork and hunches is the point I am illustrating here. It is my 'creative liberty' that Cocky identified Branch 1 of N11-88, on Gorodomlya Island, and marked it with a white pin.
12. Thermionic valves were superseded by transistors in 1947. Both components amplify weak electronic signals.
13. Eric and Cocky frequently discussed the merits of inertial guidance. A major sticking point in taking it seriously as a design option was the sheer size of the analogue computer that would be needed. There was no such thing as a 'tiny' one that could fit inside a ball, at this time.

14. The discussion about the word 'lucky' is partly true, although the specifics about what people thought it might refer to are lost in time.
15. One of the principles of the listeners was to remember that humans had invented the codes, and that they often used quite human concepts, such as types of food.
16. Russian defectors from Stalin's regime were not unusual. This may have been a former prisoner of war, useful for his native tongue, and heavily vetted to ensure loyalty to the Allies.
17. Radio Direction Finding (RDF) was a key tool of signals intelligence. Mechanically rotated antennas were quickly superseded by electronic versions.
18. Authorised licence plates on SOXMIS (and the British equivalent, BRIXMIS) vehicles allowed recording and regulating of 'mission' travel within zones. The arrangement ended in 1990.
19. A married woman had no legal rights until the end of the twentieth century. She was not equal to her husband and therefore was viewed as his property.
20. Nowadays, we understand that victims of domestic violence come to believe that they are as worthless as their partner tells them. They internalise the oppression, and think they deserve the abuse, becoming self-critical and losing perspective on where fault really lies. My father and his siblings always had to work around Edith's low self-esteem.
21. Percy/Iggy never served on the front line.

Chapter 4: Gorodomlya Island

1. The Gröttrups shared this radio set, bought in Moscow, with about five other families.
2. There was a domestic helper, but this was late at night and Irmgard was being hospitable.
3. Helmut Gröttrup's specialism was high frequency (radio wave) electronics, but the high frequency lab on Gorodomlya had not succeeded with missile guidance systems. (CIA Report 80)
4. Helmut had wanted V.I. Kuznetsov, who appears later in the book.
5. Fritz Karl Preikschat managed the 'high frequency lab' under Gröttrup from 1946 to June 1952, when he was released. As all released scientists did, he briefed the US Air Force and CIA.
6. I have not been able to discover what RM stands for.
7. How my father quickly lost his moral indignation about the uses for which these rockets were being built is fascinating to me. Everyone knew they were working on deadly missiles. He said, whether or not he risked his life by not

co-operating, there would always be another scientist ready to assist – possibly a more experienced and appropriate one. He reckoned he could slow things down through genuine ignorance, making just enough progress not to be taken off the project. The last thing he wanted was an accurately guidable rocket with a nuclear warhead attached in the hands of notorious hothead Stalin.
8. In Irmgard's journal *Rocket Wife* in March 1951, she writes: 'Moscow sends almost daily requests for further particulars about the R-14.' By September of that year: 'Inquiries which keep on coming in from commissions in Moscow make Helmut realise the great interest aroused by his R-14.'
9. Massachusetts Institute of Technology.
10. According to CIA Report RDP-80, by this time in the life of the work on Gorodomlya Island (summer 1952), the Russians had officially removed the task of developing radio control from the scientists on Gorodomlya. Ad hoc projects as defined by the main laboratory in Moscow were still being followed. One possibility is that Gröttrup was still trying to 'solve' radio guidance for larger rockets; another is that the project set by Moscow was to switch to inertial guidance.
11. With onboard gyroscopes sensing position, and accelerometers sensing acceleration, continuous data is sent to an analogue computer, programmed with the desired flight trajectory. When the rocket's velocity and direction is calculated to be over the target, the engine cut-out is activated.
12. Components invariably took far longer to arrive. This was another delaying tactic.
13. Helmut's anxiety to regain lost status with the Russians may stem from 1946, when he and Korolev jointly ran the Nordhausen rocket development facilities at Bleicherode in Germany. They were equals there – Helmut even had his own *Buro Gröttrup* – but now Gröttrup is clearly subordinate to Korolev. Yet, he is still supplying designs and technical knowledge, particularly for the R-14.
14. A session of the USSR Academy of Sciences in Leningrad in January 1949 calls for struggle against 'kowtowing to the West' and for the 'primacy of Russian Science'.
15. Irmgard says in *Rocket Wife*: 'All of a sudden the Minister embraces Director K and Director K, in his turn, embraces Helmut.'
16. Real name not recalled.

Chapter 5: Escape from Gorodomlya, Spring 1953

1. The existence of Soviet informers living amongst the community of Gorodomlya was suspected.

2. Beginning in 1948, the Communist Party organised active searches for Russians to be credited as originators of all inventions, discoveries and theories, past and present. No foreigner could be credited. Kuznetsov was a respected Russian scientist, but because of this stance, the question of whether he genuinely invented the inertial guidance system can't be ascertained.
3. The Soviet Secret Police, a powerful organisation; NKVD until 1954, when the name changed to KGB.
4. This is anachronistic, as the meme 'It's not rocket science' wasn't used until much later.

Chapter 6: Back to the UK, 1953

1. Baikonur is now in Kazakhstan. It was a remote site chosen to accommodate the radio guidance system of the R7 rocket. The R7 carried a warhead with 4–5 times the destructive power of the Hiroshima bomb.
2. The first Marriage Bureau in the UK, run by Heather Jenner and Mary Oliver, opened in April 1939, in Bond Street, London.
3. The International Geophysical Year was an international scientific project from July 1957 to December 1958. Sixty-seven countries participated, including the Soviet Union.
4. The dog, Laika, was a stray mongrel from the streets of Moscow, who died aboard Sputnik 2 in November 1957.
5. Intercontinental Ballistic Missile. The R-7 was an early ICBM. Terminology changed to ICBM for a range over 5,500km sometime before 1960.
6. The R7, the first ICBM rocket in the world, was launched from Baikonur in August 1957.
7. A few years after Baikonur was built, when Khrushchev was in power after Stalin's death, Korolev informed Khrushchev that it would be possible to dispense with radar and radio control stations. Khrushchev was dumbfounded – why then had they built Baikonur?
8. Eric held Christian beliefs.
9. Kuznetsov's gyroscopic system proved to be the most universal, and was used on the majority of Soviet rockets and spacecraft.
10. The Soviets did not include sensitive military sites on their official maps. Irmgard Gröttrup reports that when she tried to find Gorodomlya Island on a map on her release in November 1953, its location was obscured by 'brown and black blobs'.

Sources

Albring, Werner, 2016. *Gorodomlya Island: German Rocket Scientists in Russia*. Books on Demand. Edited, abridged and translated by Ursula Kuhlmann-Walter.

Aldrich, Richard J., 2001. *The Hidden Hand: Britain, America and Cold War Secret Intelligence*. John Murray.

Aldrich, Richard J. & Coleman, Michael, 1989. 'The Cold War, the JIC and British Signals Intelligence, 1948'. Taylor & Francis online.

Chertok, Boris, 2005. *Rockets and People (Vol. I)*. Contribution by Asif A. Siddiqi. NASA History Series.

Chertok, Boris, 2005. *Rockets and People (Vol. II): Creating a Rocket Industry*. NASA History Series.

Gayford, Martin, 1993. *The Best of Jazz: The Essential Guide*. Orion Books Ltd.

Gröttrup, Irmgard, 1958. *The Possessed and the Powerful: in the Shadow of the Red Rocket*. Stuttgart: Steingrueber.

Gröttrup, Irmgard, 1959. *Rocket Wife: An Account of the Enforced Sojourn in Russia of German Rocket Scientists' Families*. André Deutsch.

Gröttrup, Ursula, 14 January 2017. '*Wissenschaftliches arbeiten in einer Diktatur*'. Lecture notes from honorary speech at conference 'Helmut Gröttrup: from Rockets to chip cards'. Deutsches Museum, Munich.

Halson, Penrose, 2016. *Marriages are made in Bond Street*. London: Macmillan.

Halson, Penrose, 2017. *The Marriage Bureau*. William Morrow & Company.

Jackson, Peter, 'Cold War Warriors: Wing Commander Eric Ackermann. A Hero in the Shadows'. *Emmitsburg News-Journal*, emmitsburg.net.

Jackson, Peter & Haysom, David, 2022. *Covert Radar and Signals Interception: The Secret Career of Eric Ackermann*. Pen and Sword Books Ltd.

Jones, Reginald V., 1978. *Most Secret War*. Penguin Books Limited.

Jones, Reginald V., 1989. *Reflections on Intelligence*. Mandarin.

King, A.D., 1998. 'Inertial Navigation – Forty Years of Evolution'. *General Electric Company Review*. Vol. 13, No. 3.

Klee, E. & Merk, O., 1965. *The Birth of the Missile*. Harrap.

Kokonos, Lance & Johnson, Ian Ona, 2019. 'The Forgotten Rocketeers: German Scientists in the Soviet Union, 1945–1959'. *Texas National Security Review*. warontherocks.com

Mackenzie, Donald A., 1993. *Inventing Accuracy: A Historical Sociology of Nuclear Missile Guidance*. Cambridge, Massachussets: MIT Press.

Magnus, Kurt, 1971. *Kreisel: Theorie und Anwendungen (Gyroscope: Theory and Application)*. Springer Verlag, Berlin.

O'Daly, Kevin, 2016. 'Living in the Shadow: Britain and the USSR's Nuclear Weapon Delivery Systems 1949–62'. Unpublished PhD thesis, University of Westminster. https://doi.org/10.34737/qOvq7.

Oliphant, Roland, 20 January 2024. 'The Saturday Interview with George Robertson, Former Head of Nato'. *Daily Telegraph* article.

Russell, William T., 7 June 2012. 'Inertial Guidance for Rocket-Propelled Missiles'. *Aerospace Research Central (ARC)*, Vol. 28, Issue 1. Published online. arc.aiaa.org

Schlegel, Kristian, 1981. 'Ionospheric Research in Germany Prior to Karl Rawer'. *Advances in Radio Science*. www.adv-radio-sci.net/12/225/.

Siddiqi, Asif A., 2009. 'Germans in Russia: Cold War, Technology Transfer, and National Identity'. *Osiris* Vol. 24. Science and National Identity series 2. University of Chicago Press.

Tomayko, James E., 1985. 'Helmut Hoelzer's Fully Electronic Analog Computer'. *IEEE Annals of the History of Computing,* Vol. 7 (3): 227–40.

Wilson, G.D. (Squadron Leader), 1971. *History of Gatow 1933–1972*. Unpublished.

Zak, Anatoly, September 2003. 'The Rest of the Rocket Scientists'. *Air & Space* magazine. smithsonianmag.com

Index

ACKERMANN, Eric, xv, 145, 219
  career, 15, 56, 62, 73, 104, 108, 136, 181
  family, 32, 57, 58, 62, 181, 182
ACKERMANN, Marianna, 56, 57–8, 59-62, 181–2, 219
Air Ministry:
  Conference, 180, 210
  Headquarters, 9–10, 13, 18, 20–1, 23, 56, 58, 171, 172, 182, 213, 221, 224
Air Raid Precautions (ARP), 2, 17, 18, 61
Air Scientific Research Unit (ASRU), 8, 53, 56–8, 89, 104–107
  name change to 646 Signals Unit, 173
  purpose, 20, 24, 48, 62, 64–5, 68
ALBRING, Werner, 102, 114, 115
Analogue computer, 34, 59, 75, 103, 104, 109, 113, 119, 123, 220, 222
APPLETON, Sir Edward, 13–15, 95, 219, 222

Baikonur (cosmodrome, Kazakhstan), 174, 176, 178, 181, 211
Baruch Plan:
  consequence of USSR not signing, 71
Bleicherode *see* Nordhausen
Bletchley Park *see* GCHQ
Blitz, London, 16–17

BRAUN, Wernher von, 63, 104, 220
Bristol Aeroplane Company, 2, 4, 6, 10, 18, 44, 53, 80, 82, 159, 220, 224
BRIXMIS, 87, 89
Brocken, The, 88–92, 142, 156, 175, 224
Bruneval Raid, 23–4

Changi, Singapore *see* RAF Changi
CHURCHILL, Sir Winston, 8, 9, 13, 23, 134, 171–2, 212, 213, 214, 219–20
City 40/Ozersk, 79, 134, 173–4, 209, 212–15, 226
Cold War, 69, 71, 177
  consequences for people, 195
  consequences for world order, 129
Coronation of Queen Elizabeth II, 171
Cosmic static, 39, 43
COX, Bruce Neville, xv, 220
  career, 2, 5–7, 22, 43, 53, 64, 189, 217
  family, 1, 3, 36, 80–3, 144, 148, 149, 192, 196, 207, 216, 217
  nickname, disguise name, false name, xv, 11, 126, 151, 175, 182

DIEMINGER, Dr Walter, 14, 15, 27–31, 35, 37–40, 41, 42, 43–7, 48, 51–2, 59, 62–3, 95, 176, 220, 225

EASTHOUGH, Geoff, 65–76, 100
ENGELBACH, Paddy, 65–6, 74–6, 79–80, 82, 85, 86–7, 91, 92, 159, 169, 220

Finnish Winter War, 134
First rocket into space *see* Sputnik, 1
Fraunhofer Institute for Ionospheric Research, 7–8, 13–14, 20, 24–5, 27, 29, 58–9, 60, 73, 95, 98, 105, 119, 130–1, 175, 176, 220, 225

GCHQ (Government Communications Headquarters):
 location, 209
 personnel, 171
 preceded by Bletchley Park, 23
 purpose, 64, 78, 173, 209, 210, 212, 213, 215, 220
Glandular fever, 148
Goon Show, The, 65, 103, 173
Gorodomlya Island, xv, 225
 access in winter, 122, 126, 127, 136
 climate, 115
 location, 93
 purpose, 100, 102, 108, 145
GRÖTTRUP, Helmut, 94, 98, 106, 113, 116, 155, 220–1
 career, 94, 102, 109–17
 family, 96, 108, 110, 122, 148, 172
GRÖTTRUP, Irmgard, 93, 101, 105, 112, 116, 221
Gyroscopes/gyroscopic platform, 33, 34, 44, 59, 68, 103, 119

Hiroshima, 63, 69, 129

HÖLZER, Helmut, 104, 109, 222
HUNT, Dickie, 65

Inertial guidance, 34, 59, 102–105, 123, 124, 131, 137, 141, 174, 178, 210, 220, 222
Ionosphere, 14–15, 29, 37–40, 44, 46, 62, 176, 179, 219, 222

JENNER, Heather, 187–91
JONES, Reginald Vincent (R.V.), 221
 connection to Winston Churchill, xv, 8, 9, 23
 connection to Eric Ackermann and the ASRU, 8, 9–10, 14, 20, 21, 51, 53, 56, 64

Kaliningrad *see* Podlipki
Kegels (bowling/skittles), 24, 66, 74, 76, 80, 88–9
KGB, 63, 153–4, 161, 169–70, 172, 175, 209–10, 216
 brutality, 169, 170, 181, 215
 former name NKVD, 131, 223
KOROLEV, Sergei, 8, 25, 28–30, 42–5, 51, 100, 104, 109, 128, 130, 156, 180, 221
 career, 49, 52, 59, 63, 71, 72–3, 111, 115–16
 family, 28, 49, 52
 imprisonment, 49, 52, 176
 legacy, 176–7, 178–9, 211–12
KUZNETSOV, Viktor Ivanovich, 68, 70, 103, 109–10, 113, 125, 131, 141–2, 152, 174, 178, 212, 222

Leeds University Rocketry Association (LURA), xvi
Longwave radio waves, 15, 38, 39, 43, 46
Luftwaffe (German Air Force), 6, 10, 18

MAGNUS, Professor Dr Kurt, 102, 222
Marriage Bureau, The, 184–6
MISHIN, Vasily Pavlovich, 222
   possibility that this was 'Mitya', 222
'Mitya', xiv, 28–9, 31, 36–7, 40, 42–5, 52, 88, 89–90, 156–7, 175-81, 210, 211, 222

Nilov Monastery, 100, 140
Nordhausen, Institute, 29, 50, 128
   co-directors Gröttrup and Korolev, 29, 63, 95, 115
   purpose, 49, 63
   RABE institute at Bleicherode (absorbed into Nordhausen), 8, 128

Official Secrets Act, 15–16, 19–20, 66, 76, 186, 217
Operation Crossbow, 94
Operation Paperclip, 31, 50
   American initiative, 14
   copying by Stalin/USSR/Russia, xv, 31, 50, 63, 71, 73, 95, 130
Ostashkov, 100, 101, 105, 115, 122, 124, 125, 126, 129, 132, 133, 152, 169, 226

Peenemünde, xv, 14, 63, 94–5, 220
PIGGOTT, Dr William Roy, OBE, 8, 10, 14, 16, 27–31, 58, 61, 222

Podlipki:
   location of main NII-88 design bureau, 24km from Moscow, 71, 95, 102, 109
   renamed Kaliningrad (not the Baltic port), 68
   renamed Korolyov in 1996 (present name) after Sergei Korolev, 221
PREIKSCHAT, Fritz Karl, 95, 98, 102, 223

Radio control/guidance, 51, 131
   need for antennas, 159, 174, 178
   subject to interference, 39
Radio set, home-built, 99, 105–106, 108–109, 120–2, 123, 130, 136, 141–2, 146–7, 152, 172, 212
RAUTAVAARA, Einojuhani (Eino), 117, 126, 132, 133, 136, 138, 143, 158, 160, 161–2, 170, 223
Royal Air Force (RAF), 5–6, 7, 10, 11–12, 18, 19, 21, 22, 145, 171, 195, 219, 224
   RAF Changi, Singapore, 53–5, 56–7, 224–5
   RAF Gatow, West Berlin, 66, 72, 173, 212, 225
   RAF Wilmslow, 7

Seliger, Lake, 72, 97, 100, 107, 119, 225, 226
   white pin, 72, 74, 99, 100, 107
   NII-88 Branch 1, 68, 71–2, 98, 100, 102, 156, 174–5, 223, 225
Signals Intelligence, 16, 21, 23, 37, 48, 58, 62, 64–5, 67, 68, 71, 76

Sounding Rockets/RMs, 15, 33, 44, 62, 98, 143
SOXMIS, 81, 86–8, 130
SPENCER, Stanley, 205–207, 223
Sputnik 1, 51, 177, 180, 181, 210, 212, 221
STALIN, Josef, 4, 49, 51, 133–4, 137, 141, 156, 176, 178–9, 211, 221, 223
   replaced by Khrushchev, 176
   death, 178

Telemetry, 59, 104, 222
Teleprinter, 66, 109
TRUMAN, US President, 14, 71

Union of Soviet Socialist Republics (USSR)/Soviet Union, led by Russia, 25, 67, 179
Ustinov, Dmitry, 111–12, 113, 116, 223

V-1 doodlebug, 16–17, 18
V-2 *Vergeltungswaffen* (revenge weapon), xv, 7–8, 17, 18, 20, 40, 49, 51, 59, 62, 64, 70, 72, 73, 94–5, 97–8, 102, 103, 130, 220, 222
Vaalimaa (border crossing), 154–5, 157, 195, 212, 226
Violin, 146, 150, 152, 155, 157–67, 172

Walter, xv, 99–101, 105–106, 110, 112–13, 119–20, 122–4, 125, 129, 130, 131–2, 137–8, 142, 152, 180, 197, 213–15

Zones of post-war Allied-occupied Germany:
   American zone, 8, 14, 29
   British zone, 8, 14, 20, 29
   Soviet/Russian zone, 30, 49, 63, 88, 224